MATHEMATICS

Queen and Servant of Science

ERIC TEMPLE BELL 1883-1960

Eric Temple Bell was born in 1883 in Aberdeen, Scotland. He studied at the University of London, came to the United States at the beginning of the century, got his Ph.D. in mathematics from Columbia University in 1912, taught at the University of Washington, and since 1926 has been professor of mathematics at the California Institute of Technology. Dr. Bell, now an American citizen, is a former president of the Mathematical Association of America and a former vice-president of the American Association for the Advancement of Science. He has won many honors for mathematical research and is a member of the National Academy of Sciences. I count at least twenty-one books by Bell, besides numerous mathematical articles . . . Bell is a lively, stimulating writer, inoffensively crotchety and opinionated, with a good sense of historical circumstance, a fine impatience with humbug, a sound grasp of the entire mathematical scene, and a gift for clear and orderly explanation.

James R. Newman in *The World of Mathematics*

These words were written before Bell became professor emeritus at Caltech in 1953. He continued to live at his home in Pasadena near Caltech until shortly before his death.

MATHEMATICS
Queen and Servant of Science

---

by **E. T. BELL**

MAA SPECTRUM

PUBLISHED BY THE
MATHEMATICAL ASSOCIATION OF AMERICA

Library of Congress Catalog Card Number: 87-062937
ISBN-0-88385-447-3

PUBLISHED IN WASHINGTON, D.C. BY
THE MATHEMATICAL ASSOCIATION OF AMERICA

MANUFACTURED IN THE UNITED STATES OF AMERICA

❖❖

Foreword

Eric Temple Bell (1883-1960), born in Aberdeen, Scotland, was an American mathematician who, like Lewis Carroll, led a curious double life. As a professional he taught at Caltech, published technical books and papers, produced a splendid history (*The Development of Mathematics*), and wrote popular expository books on mathematics. His most creative work was in number theory. The Bell numbers, a remarkable infinite set of numbers, were named in tribute to his pioneering work on them.

Bell's other mental life, a life of wild romance and fantasy, was lived under the alias of John Taine, a prolific scribbler of science fiction. About half his novels were published first as books, beginning with *The Purple Sapphire* (1924), a lost-world fantasy, and ending with *G.O.G.666* (1954), about a friendly monster. In the intervening thirty years he wrote scores of short, now-forgotten tales that ran in science-fiction pulps, many of them dealing with horrendous evolutionary mutations.

As a popularizer of mathematics Bell was unrivaled. In spite of its exaggerations and occasional historical errors, I can still recall the excitement with which as a youth I first read his best known work, *Men of Mathematics*. I devoured and digested the two small books he put together to make the volume you now hold. When it first came out in 1951, an enthusiastic review in the Pasadena *Star-News* closed with these words: "The last flap of the jacket says Bell 'is perhaps mathematics' greatest interpreter.' Knowing the author well, the reviewer agrees." Who was the generous reviewer? John Taine!

It is a tribute to Bell's perception and range of interest

v

that he covers many topics on which he anticipated new developments, although he could not, of course, know exactly what they would be. Thanks to computers, the large prime mentioned on page 10 now looks like a pygmy. In 1985 the thirtieth Mersenne prime, $2^{216091} - 1$, was found. It has 65,050 digits. (Bell discusses Mersenne primes on page 229.) The notorious four-color map conjecture (pages 150-152) has been proved with the help of monstrous computer printouts. Enormous progress has been made in the classification of finite groups (page 175), although it is still true, as Bell says, that only the Lord knows how many there are.

Writing in the early days of computers, Bell tended to see them (pages 247-251) as little more than faster machines for relieving the drudgery of big calculations. Were he living today he would surely modify that view. Computers have done far more than perform rapid arithmetic. They have opened doors to research that could not have been opened without them, and their algorithmic approach has resulted in a radically different way of looking at much of mathematics.

On pages 270 and 271 Bell struggles to answer a profound question: How is it that abstract patterns, created in the minds of mathematicians, so beautifully mesh with the physical structure of the universe? Could it be there *is* no reality other than our shifting mental impressions? Because mathematicians classify and organize those impressions, Bell reasons, it is hardly surprising that their patterns fit the scenery that gave rise to them.

Not many mathematicians will buy Bell's subjective answer. With few exceptions they take for granted, as do most physicists, the independence of nature from human minds. The riddle of the magic coincidence remains. Bell cites Einstein as one who seemed to accept the identity of sense impressions and "reality." Not so; Einstein was an unabashed realist. His opposition to quantum mechanics sprang mainly from his unease with its solipsistic tendencies; from his unshakable conviction that the moon's existence doesn't require a mouse to observe it.

It is hard to improve on James Newman's description of Bell in the first of his four-volume anthology *The World of Mathematics*: "lively, stimulating, inoffensively crotchety and opinionated, with a good sense of historical circumstance, a fine impatience with humbug, a sound grasp of the entire mathematical scene, and a gift for clear and orderly explanation." *Mathematics: Queen and Servant of Science* is Bell at his popularizing best. It continues to be one of the finest of all introductions to the rich diversity of those fantastic structures that mathematicians invent, explore, and apply with such mysterious success to the hugh unfathomable world outside the little organic computers at the top of their heads.

Readers interested in learning about the recent developments mentioned earlier in the Foreword will find the following references useful. The thirtieth Mersenne prime was the topic of an article by Lee Dembart "Prime Number Found Is the Largest Known" in the *Los Angeles Times*, September 17, 1985. On Mersenne primes in general, see Chapter 3, in Beiler's *Recreations in the Theory of Numbers* (Dover, 1964).

The two mathematicians who proved the four-color conjecture, Kenneth Appel and Wolfgang Haken, wrote two non-technical articles about their proof: "A Solution of the Four-Color Theorem" in *Scientific American*, October 1977 (pages 108-121) and "The Four Color Theorem Suffices" in *The Mathematical Intelligencer*, Volume 8, 1986 (pages 10-20).

A giant step in the classification of finite groups was

taken in 1985 when all the finite simple groups were finally enumerated. See "The Enormous Theorem" by Daniel Gorenstein in *Scientific American*, December 1985 (pages 104-115), and "Ten Thousand Pages to Prove Simplicity" by Mark Cartwright in *The New Scientist*, Volume 106, May 30, 1985 (pages 26-30). Earlier articles about this search are: "Finite Simple Groups" by James Hurley and Arunas Rudvalis in *American Mathematical Monthly*, Volume 84, November 1977 (pages 693-714), "The Search for Finite Simple Groups" by Joseph A. Gallian in *Mathematics Magazine*, Volume 49, September 1976 (pages 163-176), and "Groups and Symmetry" by Jonathan Alperin in *Mathematics Today* edited by Lynn Arthur Steen (Springer-Verlag, 1978).

Those who wish to find out more about the Bell numbers referred to in this Foreword could start with "Mathematical Games" in *Scientific American*, May 1978 (pages 24-30).

MARTIN GARDNER

To the Reader

Those who wish to get on to the narrative may skip this introduction and pass at once to Chapter 1. But I thought some might wish to know what to expect and what not to expect.

This book is a thorough revision and a very considerable amplification of two popular accounts of mathematics, *The Queen of the Sciences*, 1931, written originally for the Century of Progress Series in connection with the Chicago Century of Progress Exposition, 1933, and its sequel, *The Handmaiden of the Sciences*, 1937. Instead of attempting to meet requests for reissues of each, I have largely rewritten both in combining them into one book.

The material is about equally divided between pure and applied mathematics. The two are inseparable.

The chapters need not be read consecutively nor need any chapter be read clear through unless it interests you. Take what you want and let the rest go. If something seems (and may be) too hard or unattractive at a first glance, skip it, and come back to it if you are still interested. As the mathematician J. le R. d'Alembert (1717–1783) advised, "Go on, and faith will come to you."

If you should wish to follow some topic through more than one chapter, please note the bracketed decimal numbering of sections in references. This device provides for cross references and does away with an inadequate or confusingly detailed index. For example, [3.4] refers to Chapter 3, Section 4. The mathematician and logician G. Peano (1858–1932) invented essentially this scheme in the 1890s. It is now widely used in technical writing.

A gratifying thing about the former books was the large

ix

number of nonmathematicians—lawyers, doctors, engineers, businessmen, writers—who remembered enough of their elementary mathematics to be curious about what lies beyond, and who had the interest to seek the spirit of it all without attempting to anatomize the gross body. An even greater satisfaction was that numerous young students, both boys and girls, caught a glimpse of mathematics beyond what they had to learn in school. If this book has some of the same sort of appeal, I shall be grateful to my readers.

I trust that my account of selected topics from the huge accumulation available will at least suggest that mathematics is vigorously alive and still growing, and that it is indispensable for an understanding of some sciences and technologies. It is also suggestive, I believe, for certain of the deeper parts of the philosophy of science.

My illustrations from the sciences have been drawn mostly, but not entirely, from physics and astronomy, because these can be described without long digressions. Few who are not mathematicians by trade realize how deep and wide the mathematical way of thinking is. Within the restricted frame of a sketch like this I can outline only a few examples of the prophetic insight of mathematics when directed to science by a master. As H. R. Hertz (1857–1894), the first (1888) to demonstrate experimentally the existence of wireless waves, once remarked, "It seems as if the mathematical implements we use are wiser than we, and perform their evolutions independently of our will."

Limitations of space have precluded many astonishing triumphs of mathematical divination, to make possible sufficiently detailed explanations of accepted and apparently enduring contributions of all mathematical thinking to the stock of reliable knowledge. For mathematics, whatever the threat of its continuing internal conflicts [20.1–20.4], persists.

Please don't expect more than the little offered. My sketch is not intended to be a substitute for a textbook or treatise on any subject in either pure or applied mathematics. My aims are far more modest. I trust that some mathematically inclined student may ask himself, as Descartes (1596–1650) did at the age of twenty-three as the consequence of a dream, "What way of life shall I follow?" He then went ahead and invented analytic geometry, one of the greatest mathematical achievements of all history. I also trust that mathematical amateurs, taking whatever they may like from the narrative, will sense enough of the spirit of modern mathematics to make them want to go on to fuller accounts than mine.

If occasionally I recall classical results that probably have been familiar to the reader since the first year of high school, it is because I am old-fashioned enough to believe that not all the great mathematicians of the past are dead. The latest novelty in either science or mathematics may lose its charm and freshness in a year or less. For, as any woman will tell us, today's new bonnet is tomorrow's old hat—or likely to be. So I shall continue in my belief that brains like those of Pythagoras, Archimedes, Newton, Leibniz, Gauss, Lobachewsky, Galois, Riemann, Maxwell, and Einstein are unlikely to be superseded (in my time) by even the most ingeniously devised calculating machine.

Following the table of contents there is a list of mathematicians with their initials and dates and the numbers of the sections in which they are first mentioned. This will be helpful in going through chapters and keeping an eye on epochs. Once a man has been cited, neither his initials nor his dates are given again in the text unless there is a special reason. Dates for a few were unobtainable. When I had compiled the list I was disconcerted at the number (202) of names cited. All had been recalled casually from

memory. If some mathematical reader misses his hero, may
I remind him that many of the citations were almost
fortuitous, and that to do justice to all the creators of
mathematics I should have had to include about 8,000
names? Further, the mere mention of a man's name is no
indication of what he did, either in interest or bulk. To
cite a few examples of the latter, the collected works of
Cayley [1.1] fill some 7,500 quarto pages; those of Sylvester
[1.1], about 2,800; of Abel [1.3], about 950; of Cauchy
[5.3], over 11,000; while Euler [1.3], whose works are still
being collected, undoubtedly will surpass all of these. It
has been estimated that 100 quarto volumes of about 500
pages each will be required. The contributions of several
other mathematicians not mentioned are equally impressive.
But as I am not writing a history of mathematics, I have
not attempted to name even all the major mathematicians
responsible for what mathematics has become since its
far beginnings in Sumeria, Babylonia, Egypt, and ancient
Greece. However, among the mathematicians and scientists
mentioned are some, for instance James Clerk Maxwell
(1831–1879), who influenced the course of civilization as
few if any statesmen or soldiers have yet influenced it.
How many persons on a random poll would know who
Clerk Maxwell was, or what a direct consequence of his
work did to save Britain from defeat early in World War
II? I leave it up to you.

Before proceeding to the narrative I wish to thank The
Williams & Wilkins Company of Baltimore, the original
publishers of the books which suggested this, for their
kind permission to make whatever use of the two I wished.

ERIC TEMPLE BELL

Contents

List of Names

Abel, N. H. (1802–1829)   [1.3]
Adams, J. C. (1819–1892)   [13.4]
Aesop (sixth century B.C.)   [8.9]
Airy, G. B. (1801–1892)   [13.4]
Alexander the Great (356–323 B.C.)   [11.1]
Alexander, J. W. (1888–)   [8.7]
Antipho (fifth century B.C.)   [14.4]
Apollonius (260?–200? B.C.)   [13.2]
Archimedes (287–212 B.C.)   [1.2]

Babbage, C. (1792–1871)   [11.8]
Bachet, C. G. de Mézirac (1581–1638)   [11.4]
Bachmann, P. (1837–1920)   [11.3]
Banach, S. (1892–1941)   [8.9]
Baravalle, H. (contemporary)   [10.1]
Barrow, I. (1630–1677)   [19.2]
Berkeley, G. (1685–1753)   [20.1]
Birkhoff, G. (1911–)   [6.1]
Birkhoff, G. D. (1884–1944)   [6.1]
Bohr, N. (1885–)   [10.3]
Boole, G. (1815–1864)   [5.1]
Brahe, T. (1546–1601)   [13.2]
Brouwer, L. E. J. (1881–)   [1.3]
Bryso (fifth century B.C.)   [14.4]
Burali-Forti, C. (1861–1931)   [19.4]
Butler, S. (1835–1902)   [1.1]

MATHEMATICS

Queen and Servant of Science

❖❖❖❖❖❖❖❖❖❖❖❖❖❖❖❖❖❖❖❖❖❖❖❖❖❖❖❖❖❖❖

Chapter 1

POINTS OF VIEW

❖ *1.1* ❖ *The Object of Mathematics*

Mathematics is **Queen** of the Sciences and Arithmetic the Queen of Mathematics. She often condescends to render service to astronomy and other natural sciences, but under all circumstances the first place is her due.

So said the master mathematician, astronomer, and physicist C. F. Gauss (1777–1855). Whether as history or prophecy, Gauss's declaration is far from an overstatement. Time after time in the nineteenth and twentieth centuries, major scientific theories have come into being only because the very ideas in terms of which the theories have meaning were created by mathematicians years, or decades, or even centuries before anyone foresaw possible applications to science.

Without the geometry of G. F. B. Riemann (1826–1866) invented in 1854, or without the theory of invariance developed by the mathematicians A. Cayley (1821–1895), J. J. Sylvester (1814–1897), and a host of their followers, the general theory of relativity and gravitation of A. Einstein (1878–) in 1916 could not have been stated. Without the whole mathematical theory of boundary-value problems—to use a technical term which need not be explained now—originating with J. C. F. Sturm (1803–1855) and J. Liouville (1809–1882) in the 1830s, the far-reaching wave mechanics of the atom developed since 1925 would have been impossible.

The revolution in modern physics which began with the

1

work of W. Heisenberg (1901–) and P. A. M. Dirac (1902–) in 1925 could never have started without the necessary mathematics of matrices invented by Cayley in 1858, and elaborated by a small army of mathematicians from then to the present time.

The concept of invariance, of that which remains unchanged in the ceaseless flux of nature, permeates modern physics, and it originated in the eighteenth century in the purely arithmetical work of J. L. Lagrange (1736–1813).

These are but a few of many similar instances. In none of the scores of anticipations of fruitful applications to science was there any thought of what might come out of the pure mathematics. Guided only by their feeling for symmetry, simplicity, and generality, and an indefinable sense of the fitness of things, creative mathematicians now as in the past frequently are inspired by the art of mathematics rather than by any prospect of immediate utility in the sciences and technologies.

But art for art's sake does not tell the whole story. An old example is the classical theory of heat conduction initiated by J. B. J. Fourier (1758–1830) in 1822, which led to the vast and ever-expanding theory of Fourier analysis. This is some of the purest of pure mathematics, and it continues to absorb the main efforts of scores of mathematicians who have little if any interest in applied mathematics. In applied mathematics Fourier analysis today has a far wider scope than Fourier ever imagined. It is indispensable, for instance, in all physics where wave motion underlies the pattern of events.

There are numerous other instances of this 'right-hand, left-hand' aspect of mathematics. The pure serves the applied, the applied pays for the service with an abundance of new problems that may occupy the pure for generations. The debt may then be reversed, when art for art's sake pays off in the solution of difficult problems in science and technology.

A modern instance is what happened after 1938. World War II made practical demands of such difficulty on mathematics that mathematicians had to stretch themselves to supply what was urgently needed. Mathematics as it existed in 1938 was not always adequate for the problems of the succeeding seven years. Exact solutions of critical and extremely difficult problems were out of the question in emergencies, and methods of sufficiently accurate approximation had to be devised to produce usable solutions. These methods in turn reacted on pure mathematics in the form of problems which, until military necessity demanded them, had not attracted mathematicians gratifying their personal tastes for symmetry and other manifestations of mathematical beauty. The numerical computations in some of the problems far exceeded human capacity, and new types of calculating machines had to be invented and built to do the inhuman arithmetic. In some departments of mathematics more was discovered in less than a decade of war than might have been found in half a century of peace. I believe any mathematician who has an inkling of what was accomplished in non-linear differential equations and in non-linear mechanics will agree. With the cold war of the delayed peace, some of these superhuman machines were released here and there for a few hours now and then to attack problems in the purest of pure mathematics, such as the theory of numbers. They did in a matter of minutes or hours what no mortal could have achieved in a superhumanly long lifetime.

This passage from the practical to the not immediately practical is a historical echo of a similar progress during and shortly after the Napoleonic Wars. It would be interesting to debate which one—war or economic necessity—has been the more influential in the development of mathematics, but this is not the occasion to do it. My own opinion is that war has outranked economics at least two to one. As this is written much research in pure mathematics is being financed by the military.

Another historical echo is of ironic interest. In the nineteenth century, 'the mind' was mechanically dissipated in the crude steam-engine and energy analogies then fashionable. The echo of all that furious nonsense has bounded back, amplified but recognizable, in the assertion that the human nervous system, including the 'thinking brain,' mimics an electronic supergadget—not the other way about; the gadget does not imitate the nervous system. The gadget preceded the brain, at least in the realm of Platonic Ideas.

The lightning calculations of the machines were so spectacular, when contrasted with the plodding arithmetic of the unaided human nervous system, that some of the more visionary prophets all but went overboard in their scanning of the distant horizon of the future. In the spring of 1949, for instance, experts in numerical analysis from all over the United States assembled for a solemn conference on 'thinking machines' and their possibilities for good and evil. One expert foresaw the thinking machines reproducing themselves like bacteria and crowding humanity off the globe. Another agreed, but tempered the prophecy of disaster with some common sense, leaving Homo Sap this shred of dignity or conceit: to reproduce the 'thinking' of even a very ordinary professor of mathematics, the electronic gadgetry required would smother the North American continent. As for the Queen of the Sciences, a duplication of her efforts that directed the design of the machines that may dethrone her would pack intergalactic space solid with tubes and the like out to the limits of the 200-inch telescope, say a sphere a thousand million light years in diameter. All this and more of the like, attentively listened to and respectfully received, showed how far we have come from the pessimistic 1890s, when S. Butler (1835–1902), in his acid satire *Darwin among the machines*, saw human beings evolving into a race of parasites crawling over and servicing ever bigger and more intricate machines. The thinking machines in the new dispensation will do the crawling.

The machines will have their sober say later. It seems reasonably safe to assume that until World War III, when both the Queen and her machines will probably become eternally obsolete, mathematicians may continue to cultivate their gardens and do mathematics in the reasonable expectation that their labors will not be entirely superseded by unimaginative machinery for as long as the human race refrains from suicide.

The Queen of the Sciences, however, needs no shabby apology as an introduction. C. G. J. Jacobi (1804–1851) in his retort to Fourier fittingly expressed what many believe to be the true purpose of mathematics. To appreciate this we must recall that Jacobi's deliberate contributions to applied mathematics, particularly mechanics, were comparable to Fourier's involuntary contributions to pure mathematics in his theory of heat conduction. Fourier had reproached Jacobi for "trifling with pure mathematics," particularly the theory of numbers. Jacobi replied that a scientist of Fourier's caliber should know that the true end of mathematics is the greater glory of the human mind.

Although Jacobi's retort is understandable and a dignified introduction to the Queen, we shall see at the end of our whole story that 'the human mind' as manifested in mathematics has inherent limitations of which Jacobi's generation was happily unaware. It is one of the Queen's most con-

spicuous merits that she herself has exposed her weaknesses
and limitations. She is no absolute despot over a race of
spiritless and obedient slaves. Her decrees are not promul-
gated for acceptance "everywhere, always, and by all."
She too is human and the servant of her willing subjects.

❖ *1.2* ❖ *A Golden Age*

In the early nineteenth century mathematics entered its
golden age. This most prolific period in the history of mathematics had well started by 1830; the end is not yet in sight,
in spite of wars and rumors of wars. No previous age approaches this period for the depth and tremendous sweep
of its mathematics. The only other epochs at all comparable
with it are those of Archimedes (287–212 B.C.) and I. Newton (1642–1727), and these can be compared with the
nineteenth and twentieth centuries only when generous
allowance is made for the difficulties of pioneering. The
mathematical inheritance of the nineteenth century from its
predecessors was great, both in quantity and quality, so
great indeed that one prophet in 1830 lamented that "the
golden age of mathematical literature is undoubtedly past."
That splendid inheritance of at least twenty centuries had
been increased many times by the close of the century.

By a striking historical coincidence, exactly a hundred
years after the lament was uttered, a decisive ban limiting
thinkable but unattainable goals for all mathematical and
deductive reasoning was discovered. This great result, published in 1931, would have disconcerted the logicians no
less than the mathematicians of 1830 even more profoundly
than it did their successors of 1930. Some of what seemed
trivially evident in 1930 is now merely false. We shall recur
to this in Chapter 20.

So vast has been the increase of mathematical knowledge
since the early 1800s that few men would presume to claim

more than an amateur's acquaintance with more than one
of the four major divisions of modern mathematics. The
field of higher arithmetic alone is probably beyond the complete mastery of any two men, while geometry, algebra, and
analysis, especially the last, are of even greater extent. If
mathematical physics be annexed as a province of mathematics, a detailed, professional mastery of the whole domain
of modern mathematics would demand the lifelong toil of
twenty or more richly gifted men.

In all this there is a crumb of comfort for those whose
mathematical training ended with their last year in high
school or their first year in college. These are not so much
worse off, relatively, than the majority of mathematicians
who turn the pages of the current mathematical periodi-
cals or attend scientific meetings. Out of fifty mathematical
papers presented in brief at such a meeting, it is a rare
mathematician indeed who really understands what more
than half a dozen are about. The very language in which
most of the other forty-four are presented goes clean over
the head of the man who follows the six reports nearest his
own specialty.

Many causes contribute to this state of affairs, which
seems to be a necessary consequence of mathematical
progress. I need mention only one. It is the perennial youth-
fulness of mathematics itself which marks it off with a dis-
concerting immortality from the sciences.

In theoretical physics it is but seldom necessary to master
in detail a work published over thirty years ago, or even to
remember that such a work was ever written. But in mathematics the man who is ignorant of what Pythagoras said in
Croton in 500 B.C. about the square on the longest side of a
right-angled triangle, or who forgets what someone in China
proved last week about inequalities, is likely to be lost. The
whole terrific mass of well-established mathematics, from

the ancient Babylonians to the modern Chinese and Japanese, is as good today as it ever was.

Looking down and far out over the past from our vantage points of today, we can only marvel at the dogged courage and persistence of the explorers who first won a devious way through the wilderness. Broad highways now cross the barren deserts, straight as taut strings; where scores perished miserably in the pestilent marshes there is a thriving city; and the pass through the iron mountains which our forefathers sought in vain is an easy four hours' pleasure trip from the distant city. The loftier range behind the one on which we stand is now accessible to us, although the way is hard, and by scaling its lesser peaks we can catch glimpses of an El Dorado of which the most daring of the pioneers never dreamed.

If we marvel at the patience and the courage of the pioneers, we must also marvel at their persistent blindness in missing the easier ways through the wilderness and over the mountains. What human perversity made them turn east to perish in the desert, when by going west they could have marched straight through to ease and plenty? This is our question today. If there is any continuity in the progress of mathematics, the men of a hundred years hence will be asking the same question about us. We know that there is a higher range behind the one on which we stand, and we suspect that behind that one is a higher, and so on, for as far and as long as there shall be human beings with the spirit of adventure to heed the whisper of the unknown. At the present rate of progress our vantage points of today will be barely distinguishable hillocks in a boundless plain to the explorers of a century hence. Before standing on one or two of these hard-won peaks, now easily accessible, to see what we can, let us look about us well before we start.

❖ *1.3* ❖ *Abel's Advice*

To get some sort of perspective, let us consider roughly the kind of mathematics acquired by a student who takes all that is offered in a good American high school. The geometry taught is practically that of Euclid and is about 2,200 years old. It is a satisfactory first approximation to the geometry of the physical universe, and it is good enough for some engineers, but it is not that which is of vital interest in modern physics, and its interest for working mathematicians evaporated long ago. Our vision of the universe has swept far beyond the geometry of Euclid.

In algebra the case is a little better. A well-taught student will master the binomial theorem for a positive whole number exponent which Pascal (1623–1662) discovered in 1653. There he will stop. And yet the really interesting things in algebra are the creation of the nineteenth and twentieth centuries, and began to be developed over a century and a half after Pascal died.

Of the higher arithmetic—Gauss's Queen of Mathematics —the graduate of a good school will learn precisely nothing. Unless extremely fortunate, he will never even have heard of the theory of numbers. And yet at least one of its most beautiful and far-reaching truths was known to Euclid. Many of the most striking results in this field are accessible to anyone with a year of high-school training.

As the statement and proof of Euclid's theorem on primes offer a classic illustration of the last remark, I shall reproduce them here as an example of easily understood mathematics at its best on an elementary level. A number p greater than 1 is called prime if the only numbers dividing p without remainder are 1 and p itself. For example, the following numbers are primes: 2, 3, 5, 7, 11, 13, 17, 19, . . . , 101, . . . , 257, . . . , 65,537, . . . , and (the largest prime known up till 1950):

170, 141, 183, 460, 469, 231, 731, 687, 303, 715, 884, 105, 727.

How many primes are there, even if we cannot find them all? Euclid's theorem states that there is no end to the primes: beyond any given prime there are always larger primes. To prove this, Euclid said that if there are only a finite number of primes, there must be a greatest of them all, say P. For brevity I shall assume two theorems on which the proof rests: any integer has at least one prime divisor; the number of (integer) divisors of any integer is finite. The first was proved by Euclid; the second, although he does not state it explicitly, is implied by what he does prove. Imagine all the (assumed finite number of) primes multiplied together, and add 1 to the result,

$$2 \times 3 \times 5 \times 7 \times 11 \times \cdots \times P + 1.$$

Divide this number by each of the primes 2, 3, 5, . . . , P. The remainder in each instance is 1. So the above number is not divisible by any of these primes. Therefore it either is a prime itself or is divisible by some prime exceeding P. Either conclusion contradicts the assumption that P is the greatest prime. Thus 'the number of primes is infinite.' Note that the proof does not show us how to find the *next* prime after any given one such as the monster written out above—which was not found by Euclid's theorem.

Euclid's theorem is an example of what are called *existence theorems* in mathematics. Such theorems are suspect for one school of mathematical philosophers, especially where the accompanying proofs presuppose an infinity of incompletely described acts of selection. L. Kronecker (1823–1891) in the 1870s said that such proofs are meaningless, but nobody paid much attention to him till L. E. J. Brouwer (1881–) in 1907–1912 exposed the root of the trouble in an uncritical application of classical (Aristotelian) logic to infinite sets—for which it was not devised. This, however, need not perturb us now. Euclid's proof is convincing even to those who dispute its logical validity. Objectors to existence theorems ask *to be shown* a prime larger than any given prime whose existence is asserted to be proved—"Show us a prime larger than the given prime and it suffices us." This is merely the mathematical variant of an old demand which the reader may recognize as having been made first in the New Testament. So one current philosophy of mathematics may be about 2,000 years old. Skeptics, like the poor, we have always with us.

In analytical geometry and the calculus the score of the well-taught high-school student is again zero. The calculus, however, which has been estimated as the most powerful instrument ever devised for scientific thought, may become part of the regular high-school course before World War III —if there is to be a third. Since 1900, or perhaps earlier, it has been regularly taught in the scientific course of German secondary schools.

Without a good working knowledge of the differential and integral calculus created by Newton and Leibniz in the seventeenth century, it is impossible even to read serious works on the physical sciences and their applications, much less to take a step ahead. The like is true, but to a far lesser extent, for some branches of biology and psychology, and also for some economics and sociology. Any normal boy or girl of sixteen could master the calculus in half the time often devoted to stumbling through Book I of Caesar's *Gallic War*. And it does seem to some modern minds that Newton and Leibniz were more inspiring leaders than Julius Caesar and his unimaginative lieutenant, Titus Labienus.

The college student will be considerably farther ahead at the end of his second year. Provided he has not sought culture by the literary trail exclusively, he may be able to appreciate some of the minor classics of science. He will know as much as the men of the eighteenth century knew of the calculus, and he will know it better than they did. Much of what passed for proof with the pioneers would not now be tolerated in a college textbook. To this slight extent the profound critical work of the nineteenth-century mathematicians has influenced the thinking of those who take the calculus in college—at least in a good college under a man who is not hopelessly dry and dusty.

Before quitting this somewhat uninspiring prospect, let us glance at another of the reasons why the average graduate of a standard four-year college course in mathematics usually manages to miss completely the spirit of modern mathematics. The point is clear from a remark of N. H. Abel (1802–1829), one of the greatest mathematical geniuses of all time. In his wretched life of less than twenty-seven years Abel accomplished so much of the highest order that one of the leading mathematicians of the nineteenth century, C. Hermite (1822–1901), could say without exaggeration, "Abel has left mathematicians enough to keep them busy for five hundred years." Asked how he had done all this in the six or seven years of his working life, Abel replied "By studying the masters, not their pupils."

To appreciate the living spirit rather than the dry bones of mathematics, it is necessary to inspect the work of a master at first hand. Textbooks and treatises are an unavoidable evil. The mere bulk of the work to be assimilated in any reasonable time precludes intimate contact with the creators through their works. Nevertheless it is not impossible in the ordinary course of education to read at least ten or twenty pages of mathematics as it came from the pen of

a master. The very crudities of the first attack on a significant problem by a master are more illuminating than all
the pretty elegance of the standard texts which has been
won at the cost of perhaps centuries of finicky polishing.

It is rarely feasible for beginners to attempt the mastery
of recent work. This appears in the mathematical journals,
of which, until World War II, there were about 500 published throughout the world. Some came out monthly,
others quarterly, and the contents of about 200 were almost
exclusively accounts of current mathematical research.
Most of the articles were in English, Italian, French,
or German, particularly the last two. For a competence
in modern mathematics, a reading knowledge of these
four languages is a necessity. Many articles, however,
were printed in the native languages of the authors,
ranging from Japanese, Russian, and Polish to Czech and
Rumanian.

When World War II struck, many mathematical periodi-
cals suspended publication, some permanently. Beginning
in 1948 there was a marked resumption of mathematical
publication in languages, including German and Italian,
that had been in disfavor during and after the war. This
was gratifying to mathematicians, scientists, and some
others of all nations who questioned the chauvinistic in-
junction, "Hate your neighbor as you should hate your-
self." Among other things changed by the war was the belief that politicians who would govern science are competent
to direct science for the public good. Fools have always
been governed by fools and doubtless always will be, but
not all scientists and mathematicians are yet fools. Although the Queen of the Sciences served her inferiors in the
great betrayal, she still kept her independence.

Instead of trying to touch the spirit of modern mathematics through any of this current or recent work, it is

much more practicable to study attentively some older classic. Many of the fluent papers of L. Euler (1707–1783), for example, dealing with quite elementary things, may be read as easily as a detective thriller. A little farther along, a memoir by Lagrange would make an excellent companion to all the clumsy textbooks of the standard college course in analytical mechanics. Closer to the present, Cayley's paper of 1858 which started the theory of matrices (to be noted in a later chapter) is within easy reach of a beginner. In reading Lagrange, a modern student may be baffled occasionally by the absence of the current notation for partial derivatives.

In this connection there is an amusing bit of history from the 1930s. At one of the leading American universities the ambitious president had so thoroughly grasped Abel's precept about studying the masters in preference to their pupils that he proceeded to put it into effect in the freshman class. To aid him in this worthy undertaking, the president called in a specialist in the teaching of science—not a specialist in science. Between them they made up a list of mathematical classics to be read by freshmen in their spare time. These included Newton's *Principia* of 1687 and Einstein's *Theory of general relativity and gravitation* of 1916. The last is quite a short trifle. It is the famous paper of which it used to be said that only twelve men in the world could understand it. The president was enthusiastic about the project. The freshmen were not.

❖ *1.4* ❖ *The Spirit of Modern Mathematics*

None of these remarks on the antiquity of the mathematics which passes as sufficient in a liberal education today, or on Abel's sound advice to would-be mathematicians, are intended in any spirit of discouragement. Quite the reverse: by admitting that it is a waste of time for those who are not mathematicians by trade to explore the minutiae of modern mathematics, we shall agree to be content with wider vistas than would satisfy a peering professional. In fact one of the outstanding achievements of the golden age of mathematics was the discovery and exploration of loftier points of view from which many fields of mathematics, both ancient and modern, can be seen as wholes and not as rococo patchworks of dislocated special problems. The details, however, remain matters which only specialists can appreciate.

The summits from which those broader points of view may be gained today seem to us, who did not have the pain of discovering them, to be ridiculously evident. Why were these outstanding peaks not seen before? Viewing the progress of mathematics, one might almost be tempted to amend Kant's rhapsody,

> Two things I contemplate with ceaseless awe,
> The starry Heavens, and man's sense of law,

by striking out 'sense of law' and substituting for it 'stupidity.' The only thing that deters us is the moral certainty that we ourselves are as blind to what stares us in the face as our predecessors were.

A good illustration is modern abstract algebra, which came into plain view quite suddenly in the 1920s. This will be described later.

If the mathematical spirit since the mid-nineteenth century can be described in a phrase, probably *ever greater generality and ever sharper self-criticism* is as just as any. Interest in special or isolated problems steadily diminished as the nineteenth century advanced, and mathematicians became builders of vast and comprehensive systems of knowledge in which individual theorems were completely subordinated to the grander structure of inclusive theories.

The fashioning of ever more powerful weapons for the assault of whole armies of old difficulties, instead of single combat against one at a time, also characterized the golden age of mathematics. This is a cardinal distinction between mathematics as practiced today and what the greatest mathematicians did up to the first third of the seventeenth century. A tremendous increase of power, starting slowly in the eighteenth century, particularly with Lagrange in analytical mechanics, accelerated all through the nineteenth century and well into the twentieth. And through and over the whole period played an almost continuous brilliance of the most amazing inventiveness the world has ever known.

The other side of the picture is increasing rigor. The so-called obvious was repeatedly scrutinized from every angle and was frequently found to be not obvious but false. 'Obvious' is the most dangerous word in mathematics. Unless I slip I shall avoid using it without quotation marks.

❖❖

Chapter 2

MATHEMATICAL TRUTH

❖ *2.1* ❖ *Descriptions of Mathematics*

Whatever mathematics was in the early 1800s, it is certainly not today the meager shadow of itself which some dictionaries make of it. No doubt it takes courage amounting to rashness to quarrel with a standard dictionary, but mathematicians have never been conspicuous for that particular brand of cowardice which submits to the printed word merely because it is fat, black, and backed by authority. Disregarding tradition, some have even framed pithy definitions of their own, intended as improvements on those of the dictionaries.

Unfortunately no two of the definitions are in complete agreement. Each has some high light which reflects the bias of its author, and all taken together might give an impressionistic picture not utterly inadequate. To reproduce all these attempts to hobble mathematics in a neat phrase would amount to compiling a mathematical dictionary, and the work would be hopelessly out of date long before it was finished. A few examples must suffice.

The first description of mathematics as a whole which need be seriously considered is a much-quoted epigram which B. (A. W.) Russell (1872–) emitted in 1901: "Mathematics may be defined as the subject in which we never know what we are talking about, nor whether what we are saying is true."

This definition has four great merits. First, it shocks the self-conceit out of common sense. That is precisely what common sense is for, to be jarred into uncommon sense. One of the chief services which mathematics has rendered the human race is to put 'common sense' where it belongs, on the topmost shelf next to the dusty canister labeled 'discarded nonsense.'

Second, Russell's description emphasizes the entirely abstract character of mathematics.

Third, it suggests in a few words one of the major projects of mathematics since about 1890, that of reducing all mathematics and the more mature sciences to postulational form (which will be explained later), so that mathematicians, philosophers, scientists, and men of plain common sense can see exactly what it is that each of them imagines he *is* talking about.

Last, Russell's description of mathematics administers a resounding parting salute to the doddering tradition, still respected by the makers of dictionaries, that mathematics is the science of number, quantity, and measurement. These things are an important part of the material to which mathematics has been applied. But they are no more mathematics than are the paints in an artist's tubes the masterpiece he paints. They bear about the same relation to mathematics that oil and ground ochre bear to great art.

Although it is true in a highly important sense, of which examples will appear as we proceed, that we do not know what we are talking about in mathematics, there is another side to the story, which distinguishes mathematics from the elusive reasoning of some philosophers and speculative scientists. Whatever it may be that we *are* talking about in a mathematical argument, we must stick to the subject and avoid slipping new assumptions or slightly changed meanings into the things from which we start. To be certain that we have not shifted the subject of discussion in an involved and delicate mathematical argument, or to know that our initial assumptions really do contain all that we think we are talking about, is the crux of the whole matter. Time and again mathematicians have been forced to tear down elaborate structures of their own building because, like any other fallible human beings, they have overlooked some trivial defect in the foundations.

Before leaving Russell's definition, let us put two others beside it for comparison. According to B. Peirce (1809–1880), "Mathematics is the science which draws necessary conclusions." As Russell restates the same idea, "Pure mathematics consists entirely of such asseverations as that, if such a proposition is true of *anything*, then such and such another proposition is true. It is essential not to discuss whether the first proposition is really true, and not to mention what the anything is of which it is supposed to be true." Or again, "Pure Mathematics is the class of all propositions of the form p implies q, where p, q are propositions . . . ."

The evolution of this excessively abstract view of mathematics was slow, and in its matured form it is a characteristic product of mathematical activity of the twentieth century. Not all mathematicians would assent to a definition of this type. Many prefer something more concrete. And few mathematicians would accept the dogma that skill in manipulating postulates to produce what Kant called analytic judgments is sufficient for either the creation or the understanding of mathematics. Something more than impeccable logic is required in mathematics. An expert logician will not necessarily be a passable mathematician for all his skill in logic, any more than a scholarly prosodist will be a respectable poet for all his mastery of meter.

These estimates may well be enhanced by one from F. Klein (1849–1929), the leading German mathematician of the last quarter of the nineteenth century. "Mathematics in general is fundamentally the science of self-evident things." This has been reserved for the last because, although it is profoundly true, it might easily be misunderstood.

In the first place the modern critical movement has taught most mathematicians to be extremely suspicious of 'self-evident things.' In the second place it can be misleading for any mathematician to imply that complicated chains of close reasoning are either easy or avoidable from the beginning. After a problem has had its back broken by half a dozen virile pioneers, it is usually simple enough to walk up and dispatch the brute with a single well-aimed bullet. If mathematics is indeed the science of self-evident things, mathematicians are a phenomenally stupid lot to waste the tons of good paper they do in proving the fact. Mathematics is abstract and it is hard, and any assertion that it is simple is true only in a severely technical sense—that of the modern postulational method which, as a matter of fact, was exploited by Euclid. The assumptions from which mathematics starts are simple; the rest is not.

Each of the quoted attempts to define mathematics has contributed an illuminating touch to the whole picture. These, and the scores of others which have not been mentioned, illustrate the hopelessness of trying to paint a brilliant sunrise in one color. The attempt to compress the free spirit of modern mathematics into an inch in a dictionary is as futile as trying to squeeze an ever-expanding thundercloud into a pint bottle.

❖ 2.2 ❖ *The Postulational Method*

Up until the early decades of the twentieth century it was quite commonly thought that mathematics has a peculiar kind of truth not shared by other human knowledge. For example, E. Everett (1794–1865) expressed the popular conception of mathematical truth as follows: "In the pure mathematics we contemplate absolute truths, which existed in the Divine Mind before the morning stars sang together, and which will continue to exist there, when the last of their radiant host shall have fallen from heaven." It will be remembered that Everett delivered the speech of the day—lasting for hours—at Gettysburg (November 19, 1863) when Lincoln delivered his address in a matter of minutes.

Although it would be easy to match this extravagance by many as wild from later writings of those who, like Everett, were not mathematicians by profession, it must be stated emphatically that only an inordinately stupid or conceited mathematician would now hold any such inflated estimate of his trade or of the 'truths' he manufactures. One modern instance of the same sort of thing, and we shall pass on to something more profitable. The astronomer and physicist J. H. Jeans (1877–1946) declared in 1930, "the Great Architect of the Universe now begins to appear as a pure mathematician." If this high compliment or that of Everett meant anything, pure mathematicians might indeed feel proud. But the Queen of the Sciences is not susceptible to flattery.

Against all the senseless rhetoric that has been wafted like incense before the high altar of 'Mathematical Truth,' let us put the considered verdict of the last of the mathematical giants from the nineteenth century. Mathematics, according to D. Hilbert (1862–1943), is nothing more than a game played according to certain simple rules with meaningless marks on paper. I shall return to this in the concluding chapter. This is rather a comedown from the architecture of the universe, but it is the final dry flower of centuries of growth. The *meaning* of mathematics has nothing to do with the game, as such, and pure mathematicians pass outside

their proper domain when they attempt to give the marks
meanings. Without assenting to this drastic devaluation of
mathematical truth, let us see what brought it about.

The story begins in 1830 with G. Peacock (1791–1858)
and his study of elementary algebra. Peacock seems to have
been one of the first to recognize that algebraic formulas are
purely formal—empty of everything but the rules according to which they are combined. The rules in a mathematical game may be any that we please, provided only
that they do not lead to flat contradictions like '*A* is equal to
B, and *A* is not equal to *B*.' The British algebraic school,
Peacock, D. F. Gregory (1813–1844), W. R. Hamilton
(1805–1865), A. De Morgan (1806–1871), and others,
stripped elementary algebra of its inherited vagueness
and embodied it in the strict form of a set of postulates.
As these postulates are illuminating, I shall state them in
the following chapter in a modern version. Before doing so,
however, let us see what postulates are.

A postulate is merely some statement which we agree to
accept without asking for proof. A famous example is
Euclid's postulate of parallels, one form of which is this:

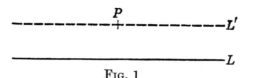

FIG. 1

"Given a point *P* in a plane and a straight line *L* not passing
through *P*, *it is assumed* that precisely one straight line *L'*
lying in the plane can be drawn through *P*, such that *L* and
L' do not meet however far they are drawn." Many geometers after Euclid's time struggled to *prove* that there *is* one
such line *L'* and, moreover, that there is *only one*. They
failed, for the sufficient reason that the postulate is *incapa-*

ble of proof, in the sense that consistent geometries can be constructed without it. I return to this in Chapters 3, 8, and 10. In passing, any modern mathematician will salute Euclid's penetrating genius for recognizing that this complicated statement about parallel lines is indeed a postulate, on a level, so far as Euclid was concerned, with such a simple postulate as 'things which are equal to the same thing are equal to one another.'

Euclid's postulate illustrates two points about postulates in general. A postulate is not necessarily 'self-evident,' nor do we ask, "Is it true?" The postulate *is given: it is to be accepted without argument*, and that is all we can say about the postulate itself. In the older books on geometry, postulates were sometimes called axioms, and it was gratuitously added that 'an axiom is a self-evident truth'—which must have puzzled many an intelligent youngster.

Modern mathematics is concerned with playing the game according to the rules; others may inquire into the 'truth' of mathematical propositions, provided they think they know what they mean.

The rules of the game are extremely simple. Once and for all the postulates are laid down. These include a statement of all the permissible moves of the 'elements'—or 'pieces.'

It is just like chess. The 'elements' in chess are the thirty-two chessmen. The postulates of chess are the statements of the moves a player can make and of what is to happen if certain other things happen. For example, a bishop can move along a diagonal; if one piece is moved to an occupied square, the other piece must be removed from the board; and so on. Only a very original philosopher would dream of asking whether a particular game of chess was 'true.' The sensible question would be, "Was the game played according to the rules?"

Among the permissible moves of the mathematical game is one which allows us to play. This is the assumption outright that the laws of ordinary logic can be applied to our other postulates. As this blanket postulate is of the highest importance, I shall illustrate its meaning with a simple example.

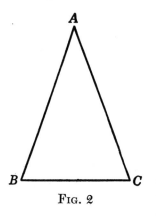

FIG. 2

In Proposition 6 of Book I of his *Elements*, Euclid undertakes to prove that *if* the angles *ABC* and *ACB* are equal in the triangle as drawn, *then* the side *AB* is equal to the side *AC*. His proof is the first recorded example of the indirect method—*reductio ad absurdum* (reduction to the absurd). Euclid provisionally assumes the *falsity* of what he wishes to prove. Namely, he assumes that *AB* and *AC* are *unequal*. This leads easily to the conclusion that the angles *ABC* and *ACB* are *not* equal. But they were *given* equal. Faced with this contradiction, Euclid concludes by common logic that his provisional assumption that *AB* and *AC* are *unequal* must be wrong. Therefore *AB* and *AC* must be equal, as this is the only way of avoiding the contradiction.

In this, when fully developed, appeal is made to two of the cardinal principles of Aristotelian logic, the law of contradiction and the law of the excluded middle. The law of contradiction asserts that no *A* is not-*A*; the law of the excluded middle asserts that everything is either *A* or not-*A*. Both of these were (almost universally) accepted until 1907–1912 in all sane reasoning, but both, be it observed, are *postulates*. As we shall see in the final chapter, the law of the excluded middle has been called into question as a universally valid part of reasoning since 1912 by mathematicians. In practically all mathematics since 1912, however, the whole machinery of common, classical logic has been included in the postulates of all mathematical systems. Unless otherwise remarked, this assumption is tacitly made in everything discussed.

Having stated a particular set of postulates, say those of elementary algebra or those of elementary geometry, what can we do next? Beginning in the 1890s and continuing well into the twentieth century, a beautiful art developed around postulate systems as things to be studied for their own sake. One question asked about a given set is this: Is the set the most economical? Or is it possible to prune off one and still have a sufficiency? If so, the one that is to be pruned must follow by the rules of logic from the others. With a little practice even amateurs can construct such desirable sets of mutually *independent* postulates. It is at least as amusing a pursuit as solving crossword puzzles or playing solitaire, and it is fully as useful as—whatever anyone cares to mention. In the next chapter I will illustrate this for the postulates of elementary algebra.

The requirement of independence for our postulate set is dictated not by necessity but by aesthetics. Art is usually considered to be not of the highest quality if the desired object is exhibited in the midst of unnecessary lumber. Many an otherwise impressive cathedral has been cheapened by too many gargoyles.

Are the postulates then completely arbitrary? They are not, and the one stringent condition they must meet has wrecked more than one promising set and the whole edifice reared upon it. *The postulates must never lead to an inconsistency.* Otherwise they are worthless. If by a rigid application of the laws of logic a set of postulates leads to a contradiction, such as '*A is B, and A is not B,*' the set must either be amended so as to avoid this contradiction (and possibly others), or it must be thrown away. We shall have blundered, and we must start all over again. At this point it is pertinent to ask, "How do we know that a particular set of postulates, say those of elementary algebra, will never lead to a contradiction?"

The answer to this disposes once and for all of the hoary myth of 'absolute truth' for the conclusions of pure mathematics. *We do not know, except in comparatively trivial instances, that a particular set of postulates is self-consistent and that it will never lead to a contradiction.* This may seem strong, but the reader will be in a position to judge for himself if he reads the succeeding chapters, particularly the last of all.

So much for the "absolute truths, which existed in the Divine Mind before the morning stars sang together"—so far as these were mathematical truths—and so much also for the Great Architect of the Universe as a pure mathematician. If He can do no better than some of the postulate systems that pure mathematicians have constructed in the past for their successors to riddle with inconsistencies, the universe is in a sorry state indeed. The less said about the postulate systems for the universe constructed by scientists, philosophers, and theologians, the better.

If anyone asks where the postulates come from in the first place, he is harder to answer. Possibly the question is of the kind which mathematicians describe as 'improperly posed.' Merely because it sounds like a sensible question is no guarantee that it is not as nonsensical as asking when time began. In passing it may be noted that the query about time seems to make sense to some astrophysicists who have tried to improve on Einstein's general relativity.

However, there have been attempts to state the question meaningfully and to answer it historically. All such attempts appeal to a hypothetical history beyond objective confirmation or refutation. Some, favored by certain anthropologists, seek to find in the behavior of contemporary primitives the tentative beginnings of abstractive thought. When projected into "the dark backward and abysm of time," the findings give a plausible enough explanation of the origin of Euclidean geometry in the sixth century B.C. At a much farther advanced stage of culture than the primitive, visual experience had suggested to Thales and Pythagoras that empirically 'true' assertions about configurations of straight lines must, in some sense deeper than mere empiricism and crude sensory experience, be 'true.' This was the hypothetical beginning of mathematics and indeed of all strict deductive reasoning.

According to Greek tradition, Thales, and after him Pythagoras, sought and found this deeper meaning of 'true' in the postulational method for plane geometry. From carefully stated *assumptions*, abstracted possibly from ocular observation but *accepted as true* without argument or further logical analysis, the first geometers proceeded to *deduce* all the logical consequences of their assumptions they could by a reasoning—common logic—that seemed necessarily innate in any rational mind. Those earliest mathematicians abstracted the data of their senses into concepts *outside* the domain of sensory experience.

By projection from this conjectured past to the present, it follows, for one school of mathematical philosophers and reconstructors of the unattainable past, that the refined mathematics of the twentieth century in its most rarefied abstractions follows the same trail from sensory 'experience' to that which is beyond recognition by the five senses. Anyone who has read some modern works on the postulational technique may be rather skeptical of the 'experience' theory, unless 'experience' includes the whims of mathematicians indulging in what has been called the decadent vice of playing with barren postulates. We shall return to this in [3.3].

Finally, one school of twentieth-century mathematical philosophers discarded 'true' in favor of 'consistent.' But then there remained the problem of proving that a particular set of postulates will not engender a contradiction. And there the matter rests.

Chapter 3

BREAKING BOUNDS

✤ 3.1 ✤ Common Algebra

A statement of what common algebra is from a modern point of view was promised in the preceding chapter. I ask the reader to look rather closely at the simple postulates given, as from them we shall see presently at least one aspect of that process of generalization which has been a distinctive feature of much mathematics since about 1900.

What follows is a paraphrase of the first part of a paper by E. V. Huntington on Definitions of a Field by Independent Postulates (*Transactions of the American Mathematical Society*, Vol. 4, pp. 31–37, 1903). The whole paper is within easy reach of anyone who can read simple formulas.

The underlying idea is that of what we call a *class* in English. The word 'set' has superseded 'class' in some departments of mathematics. We do not define class, but we do assume that, given any class, say C, and an individual, say i, we can recognize intuitively whether i is or is not a member of C. If i is a member of C, we say that i *is in* C. For example, if C is the class of horses, and i is a particular cow, we can point to i and say i *is not in* C. All this is so simple that the only difficulty is to realize that it is less simple than it seems. In fact one of the difficulties in the foundations of mathematics is to give an unobjectionable definition of 'class' or 'set.' But this need not trouble us on our present level.

To proceed with common algebra. The letters a, b, c in what follows are to be interpreted as mere *marks* without meaning. Chinese characters, or §, *, †, or any other marks would do as well. The signs ⊕, ⊙ may be given any names we please, for example, tzwgb and bgwzt. For the sake of euphony, however, they may be read plus, times.

We are *given a class* and *two rules of combination,* or *two operations,* that can be performed on any *couple* of things in the class. The operations are written ⊕, ⊙. We *postulate* or *assume* that whenever a and b are in the class, the result, written $a \oplus b$, of operating with ⊕ on the *couple a,b* is a unique thing which is in the class. This postulate is expressed by saying that *the class is closed under* ⊕. We *postulate* also that the class is closed under the operation ⊙.

A word as to the reading of formulas. Suppose a and b are in the given class. By our postulate above, $a \odot b$ is in the class, and therefore it can be combined with any c in the class to give a unique thing again in the class. How shall this last be written? If we get the result from the couple $a \oplus b$, c, we shall write it $(a \oplus b) \oplus c$; if the result is got from the couple c, $a \oplus b$, we shall write it $c \oplus (a \oplus b)$. At this step the hasty may jump to the unjustifiable conclusion that, *necessarily,*

$$(a \oplus b) \oplus c = c \oplus (a \oplus b),$$

where $=$ is the usual sign of equality.

The only things we shall assume about equality are these. If a,b are in the class, exactly one of the following is true: a is equal to b ($a = b$), or a is not equal to b ($a \neq b$).

If a is in the class, then $a = a$. This says that a thing 'is equal to' itself.

If a, b, c are in the class, and if $a = b$ and $b = c$, then $a = c$. This is Euclid's old friend about things equal to the same thing being equal to one another.

If a,b are in the class, and if $a = b$, then $b = a$.

The postulates proper for common algebra can now be

stated in short order. In this particular set there are seven, which we number for future reference.

POSTULATE (1.1). If a,b are in the class then $a \oplus b = b \oplus a$.

POSTULATE (1.2). If a, b, c are in the class, then

$$(a \oplus b) \oplus c = a \oplus (b \oplus c).$$

POSTULATE (1.3). If a,b are in the class, then there is an x in the class such that $a \oplus x = b$.

These are merely the familiar properties of algebraic addition precisely and abstractly stated. Subtraction is given by (1.3). Notice that our covering postulate of closure under $\oplus$ permits us to talk sense about $a \oplus b$ and $b \oplus a$ in (1.1), and similarly in the rest. The following three make common multiplication precise. Postulate (2.3) gives algebraic division.

POSTULATE (2.1). If a,b are in the class then $a \odot b = b \odot a$.

POSTULATE (2.2). If a, b, c are in the class then

$$(a \odot b) \odot c = a \odot (b \odot c).$$

POSTULATE (2.3). If a,b are in the class and are such that $a \oplus a$ *is not equal* to a, and $b \oplus b$ *is not equal* to b, then there is a y in the class such that $a \odot y = b$.

The seventh and last connects $\oplus$, $\odot$.

POSTULATE 7. If a, b, c are in the class, then

$$a \odot (b \oplus c) = (a \odot b) \oplus (a \odot c).$$

Notice that (1.1) and (2.1), also (1.2) and (2.2), differ only in the occurrence of the signs $\oplus$, $\odot$.

If we now replace $\oplus$ by the common $+$, and $\odot$ by $\times$, and then say that the class shall be that of all the numbers, positive or negative, whole or fractional, that ordinary arithmetic deals with, we see that our postulates merely state what every child in the seventh grade knows. *Of course*, to take (1.1), (2.1), we *must* get the same result out of $6 + 8$

as we do out of 8 + 6, and *of course* 8 × 6 is the same number as 6 × 8.

There is no 'of course' about it. Can it be *proved?* Yes, up to a certain extent, provided we *agree to stop somewhere and not demand further proof for the things asserted.* This needs elaboration.

In common algebra we point to all the numbers of common arithmetic, as we did just a moment ago, and say *there* is a class, the numbers, and *there* are two operations, common addition and multiplication, which satisfy all our seven postulates.

Examining parts of the curious (2.3), we observe that they amount to forbidding the beginner's sin of attempting to divide by zero.

If then we agree to accept common arithmetic as a self-consistent system, we shall have exhibited a consistent interpretation of our seven postulates. Otherwise, granted that arithmetic is self-consistent, we shall have pointed out a self-consistent system satisfying our postulates.

But what about common arithmetic? Why not see what *it* stands on? Do we *know* that the rules of arithmetic *can never* lead to a contradiction? No. This brusque denial will be amplified in the concluding chapter. Since Hilbert first proposed the question in 1898 a host of mathematicians have busied themselves over this. Perhaps the most striking answer is that which bases the numbers on symbolic logic. But on what is symbolic logic based? Why stop there? For the same reason, possibly, that the Hindu mythologists stopped with a turtle standing on the back of an elephant (or was it the other way about?) as the last supporter of the universe. No finality may be possible.

Another sort of answer was given by Kronecker. An arithmetician by taste, Kronecker wished to base all of mathematics on the positive whole numbers 1, 2, 3, 4, ...

His creed is summed up in the epigram, "God made the integers, all the rest is the work of man." As he said this in an after-dinner speech, perhaps he should not be held to it too strictly. Today some would say that it was not God but man who made the integers.

In the paper from which the seven postulates are transcribed, it is *proved* that the set is *independent:* no one of the seven can be deduced from the other six.

The system which the seven postulates define is called a *field*. An instance of a field is therefore the common algebra of the schools. The same system, namely, a field, can also be defined by other sets of postulates. There is not a unique set of postulates for common algebra but several, all of which have the same *abstract content*. It is just as if several men of different nationalities were to describe the same scene in their respective languages. The scene would be the same no matter what language was used.

Which of all possible equivalent sets of postulates for a field is the best? The question is not mathematical, as it introduces the elements of taste, or purpose, or value, none of which has yet been given any mathematical meaning. For some purposes a set containing a large number of postulates may be preferable. In such a set most, if not all, of the postulates will be simple subject-predicate statements. For other purposes a set in which not all the postulates are independent might be easier to handle, and so on.

Before leaving this set, let us recall that it contains *all* the rules of the game of common algebra. We can make our moves only in accordance with these rules.

We can make any rules we please in mathematics, to begin with, provided they are consistent. But, having made the rules, we must be sportsmen enough to abide by them while playing the game. If the game should prove too hard or uninteresting under the prescribed rules, we are free to make a new set and play accordingly. The exercise of that legitimate license was the source of some of the most interesting mathematics of the nineteenth and twentieth centuries. It also tempted some of the lazier mathematicians to reduce their hard problems to easy ones by adding a new postulate or two where convenient to soften things up a bit.

I have chosen algebra rather than geometry to illustrate postulate systems on account of its greater simplicity. The same sort of thing has been done repeatedly for elementary geometry, for which one of the neatest postulate systems is Hilbert's of 1899–1930.

To what has already been said about playing with postulates, may be added the following observation (1940) by H. Weyl (1885–): "I should like to point out that since the axiomatic attitude has ceased to be the pet subject of the methodologists its influence has spread from the roots to all branches of the mathematical tree."

❖ 3.2 ❖ *Changing the Rules*

To recall some useful terms, let us name the rule of play given by Postulate (1.1) the *commutative* property of the operation ⊕. As Postulate (2.1) says exactly the same thing about ⊙ that (1.1) does about ⊕, we refer to it as the commutative property of ⊙. Similarly (1.2), (2.2) express the *associative* property, and Postulate 7 is the *distributive* property. These are the familiar names of the schoolbooks on algebra.

The circles in ⊕, ⊙ can now be dropped, as they have sufficiently played their part of emphasizing that we are speaking of whatever satisfies the seven postulates and nothing else. Accordingly I shall now write $a + b$ for

$a \oplus b$, and $a \cdot b$ or ab for $a \odot b$, exactly as in any text on algebra.

Suppose now that we rub out one of the postulates, say (1.2), the associative property for addition. Then, whenever $a + (b + c)$ turns up, we can *not* put $(a + b) + c$ for it, as there is no postulate permitting us to do so. We must carry $a + (b + c)$ and $(a + b) + c$ as two distinct pieces of baggage, instead of the one piece we had before. The new algebra is more complicated than the old. Is it any less 'true'? Not at all, *provided* we can point to a class of things a, b, c, . . . and two operations, our new 'plus' and 'times,' which behave as the *six* postulates we have now laid down require, and which we agree to accept as consistent. Without bothering for the moment whether we can point to an example, let us see how the system defined by the *six* postulates compares with that defined by all *seven* from which it was derived by *suppressing one postulate*.

A moment's reflection will show that the new system is *more general*, that is, *less restricted*, than the old. This is plain, because the new system has to satisfy *fewer* conditions than the old, and therefore there is greater freedom within it. Whatever we can say about the new system will hold also for the old. The other way about is false, for *some* things (namely *all* those for which Postulate (1.2) is necessary) can *not* be said about the new.

This illustrates one way of generalizing a mathematical system. We *weaken* the postulates.

More than idle curiosity prompts the next question. By weakening the postulates of a field (common algebra) how many *consistent* systems can be manufactured? I believe the answer has not been given in the texts (it is not mine), but it appears to be at most 1,152. At any rate, mathematicians have produced well over 200 such

systems incidentally in the course of their work on postulates. There are thus 200 or more, possibly 1,151, 'algebras'
in addition to the 'common algebra' of the schools, and
each of these is *more general* than the common one. The
schoolboy of the twenty-second century may have to learn
some of these, but he certainly will not be tormented by
more than 1,152 in all, for that is the upper limit of possibilities in this direction.

Anyone except a mathematician may be pardoned for
demanding what is the good of all this. Isn't the algebra
of the high school enough for practical life? A reasonable
answer seems to be that high-school algebra is either too
much or too little for everyday life. Only one person in
hundreds ever actually uses the common algebra he learned.
But for the many in our technical age who *must* use mathematics in their work, far more than common algebra is
desirable and often necessary. Two examples must suffice
for the present to give some support to this assertion.

Open any handbook on mechanics or physics as they are
taught in the first two years of college to those who intend
to make their livings at applied science, and notice the
heavy black letters, usually in Clarendon type, in the
formulas. These represent 'vectors.' A vector is the mathematical name for a segment of a straight line which has
both length and direction. A vector **a**, interpreted physically, represents, among other things, a force of stated
amount acting in a stated direction. Now follow through a
few of the vector formulas. Presently the astonishing fact
presents itself that **a** ✕ **b** is *not* always equal to **b** ✕ **a.**

Vectors are *added* according to our Postulates (1.1), (1.2);
Postulate (2.2) is still good, and Postulate 7 is satisfied, all
with perfectly sensible physical meanings. But (2.1), the
commutative property of *multiplication*, has gone overboard,
as it *is not true for vectors.* All this, when properly amplified,

gives the standard *vector analysis,* without which no one would think nowadays of trying to master mechanics or electricity and magnetism.

Still stranger specimens of our collection have their uses. One, something like vector algebra, invented by W. K. Clifford (1845–1879) in 1872, proved of great service in studying the complicated mechanics of atoms. Others are of equal interest to mathematicians. Even the freak we suggested by suppressing (1.2) is not without charm.

An example of generalization from geometry instead of algebra will be given presently. For the moment let us glance back. All that has been said is as simple as any interesting game, and it is in fact far simpler than chess. Its simplicity did not mature overnight. Almost a century was required for the perfection of the fruit.

Hamilton, a universal genius and one of the most creative mathematicians of the golden age, racked his brains for fifteen years in the effort to create a suitable algebra for geometry, optics, mechanics, and other parts of physics. The obstacle which blocked him all those fifteen years was the commutative property of multiplication. Finally the solution flashed on him one day while he was out walking: *throw away the commutative property; a times b is not always and everywhere necessarily equal to b times a.* Today a college freshman discards the commutative property without fifteen seconds' thought.

❖ 3.3 ❖ Sources of Postulates

With a definite system of postulates now at hand for inspection, we may ask where it came from. To some mathematicians the question is meaningless. Others accept the statement of certain philosophers that the postulates of mathematical systems are derived from experience [2.2]. This may be satisfactory, provided we know what 'experience' means. But to say that every set of mathematical postulates is a fruit of experience is to stretch the meaning of 'experience' to the breaking point and to give an answer that is little better than a quibble. If indeed, as Hilbert asserted, mathematics is a meaningless game played with meaningless marks on paper, the only mathematical experience to which we can refer is the making of marks on paper.

Instead of trying to answer what may be a senseless question by giving a plausible equivocation which any competent mathematician could riddle in two seconds, let us see how one of the most celebrated systems of postulates actually originated. Anyone who wishes may ascribe the postulates already stated for a field to experience. The set for N. I. Lobachewsky's (1793–1856) geometry could more properly be credited to a lack of experience in any usual sense of the word.

For centuries before 1826, mathematicians had tried to deduce Euclid's postulate of parallels [2.2] from the remaining postulates of Euclid's geometry. They succeeded in proving that *if* the postulate is so deducible, then any one of a large number of equivalent geometrical theorems must be true. Conversely, *if* one of these theorems *is* a consequence of all of Euclid's postulates *except* the one for parallels, then it can be proved that through a point P in a plane can be drawn *exactly one* straight line L' lying in the plane determined by P and a straight line L not passing through P, such that L and L' do not meet however far extended.

One of these crucial theorems equivalent to the parallel postulate is this 'obvious' trifle. Given a segment AB of a straight line (Figure 3) and *equal* perpendiculars AC, BD erected at A and B, and on the same side of AB, join CD, and *prove* that *each* of the equal angles (they are easily proved to be equal) *ACD*, *BDC*, marked in Fig. 3, *is a right angle.*

Common sense at once 'sees' that *ACD*, *BDC* are right angles by folding the rectangle over the line perpendicular to *AB* through the middle point *M* of *AB*. What common sense thinks it sees is a striking illustration of the fact that mathematics is *not* the science of self-evident things.

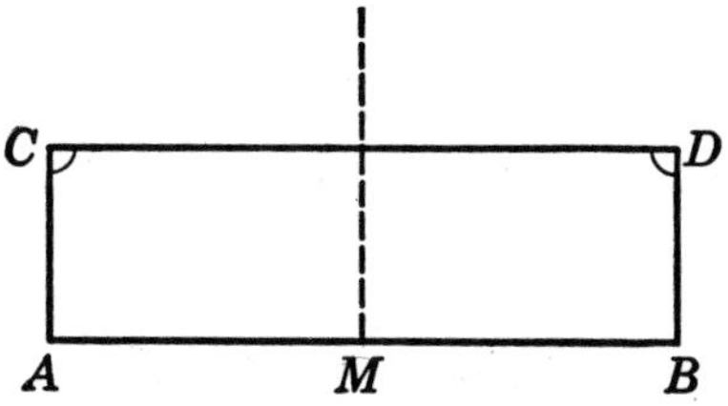

Being unable to *prove* that each of *ACD*, *BDC* is a right angle by Euclid's geometry *without* using Euclid's parallel postulate, Lobachewsky conceived the brilliant and epoch-making idea of what is equivalent to *postulating* the *assumption* that each angle is *less than* a right angle. With minute

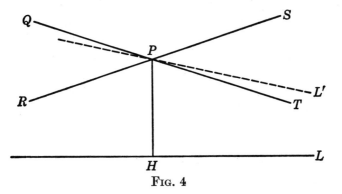

FIG. 3

FIG. 4

care he proceeded to develop the consequences of this hypothesis. It led him to a simple geometry, *just as consistent as Euclid's and equally sufficient for the needs of everyday life,* in which he discovered the following undreamed of situation regarding 'parallels.'

P is any point not on the straight line *L*; *PH* is per-pendicular to *L*; *QT* and *RS* are a particular pair of straight lines drawn through *P*. The angle *TPS* between *RS* and *QT* is *greater than zero;* that is, the lines *RS* and *QT do not coincide.* Now, in Lobachewsky's geometry, *any line L' passing through P* and lying within the angle *TPS* is such that it *never meets L,* however far extended in either direction. So then there are an infinity of 'parallels' in Lobachewsky's geometry.

In Euclid's geometry, *RS* and *QT* coincide, and there is only one parallel. Lobachewsky called the two lines *PR*, *PT*, neither of which meets *L*, his *parallels*, as they both have all the properties of Euclid's one parallel, for example, transitivity: if *A* is parallel to *B*, and *B* is parallel to *C*, then *A* is parallel to *C*.

Which geometry is 'true'? The question is improper; each is self-consistent. And each is sufficient for everyday life.

But why, out of the *three* thinkable possibilities that the equal angles *ACD*, *EDC* in the original figures are each *less than*, *equal to*, or *greater than* a right angle, stop with the first two, which give the respective geometries of Lobachewsky and Euclid? There is no compulsion. We may equally well *postulate* that each is *greater than* a right angle. The result is a third geometry, again self-consistent and sufficient for everyday life. In this last geometry (developed by Riemann) there are *no* parallels, and a straight line is *closed* and of *finite* length.

Why choose Euclid's in preference to either of the others? Some would say because Euclid's is the simplest of the three to learn, backed as it is by 2,200 years of schoolteaching.

If the common sense and experience of our immediate neighborhood suggest that Euclid's familiar geometry is the only one of any practical importance, we need only consider navigation. What follows will be of cardinal importance later when we consider some of the services mathematics has rendered the physical sciences. What is the shortest course by water from San Francisco to Yokohama? Clearly, straight across the Pacific. But what is 'straight' here? Not a straight line, as on a plane; for the straight line joining the two ports would burrow through the earth. The course is an arc of a great circle, as is plain on looking at a terrestrial globe. So we develop a geometry of shortest distances, or most direct distances, technically, *geodesics*, on a sphere. These are arcs of great circles, a great circle being the curve (a circle) in which a plane through the center of a sphere cuts the surface of the sphere.[1] These geodesics play the corresponding part on the sphere of straight lines on a plane. If the sphere be imagined as swelling indefinitely, its curvature diminishes continually, its surface in any restricted region becomes less and less distinguishable from a plane, and the arcs of great circles— the geodesics—approach closer and closer to straight lines.

Carry all this on to any surface. With sufficient accuracy for our purpose here, a geodesic joining two points P,Q, on a surface may be visualized thus. A thread is passed around pins struck into the surface at P,Q and then drawn as tight as possible. If the pins are *sufficiently close together*, to take account of humps in the surface, the stretched thread will lie entirely *on* the surface, as it does everywhere on a plane or a sphere. The thread then, by definition, follows an arc of a geodesic on the surface. This is a sensible extension of the description of a 'straight line' in a plane

---

[1] Before scientifically created rayon ruined the silk industry of Japan, great-circle sailing was an extremely practical problem: the silk had to be rushed to New York in the shortest time possible. Anyone who in the old days saw a silk train passing up the passenger limited on the Canadian Pacific Railroad will appreciate geodesics.

as 'the shortest distance between two points,' as many school geometries define a straight line. But for geodesics on a sphere there may be a significant difference from geodesics on a plane.

In a plane the stretched string gives a *unique* distance, the shortest, joining two points P,Q, and the distance *from P to Q* is the same as the distance *from Q to P*. On a sphere, with 'distance' defined as length of geodesic arc, the distance from P to Q is equal to the distance from Q to P only if P,Q are the extremities of a diameter of the sphere. Remembering always that *'distance' is to be measured along a geodesic*, we see on looking at a globe that unless P,Q are at opposite poles, one of the distances between P,Q is the longest and the other the shortest.

In the geometry of distances on a sphere there are no 'parallels,' as any two geodesics intersect in two points. Compare this with Euclid's axiom that no two 'straight' lines can enclose a space. He defined a 'straight line' as a line that lies evenly between its extremities. His geometry sufficed for a flat earth. 'In the small,' that is, in our immediate neighborhood, Euclid's geometry is adequate; 'in the large,' even when we fly only from San Francisco to New York, it is inadequate. The earth really is not flat after all; its curvature has been photographed from an airplane. This should convince doubters in the vicinity of Chicago and elsewhere.

Some of this may jar common sense. If it does, I recall Einstein's remark that "common sense is nothing more than a deposit of prejudices laid down in the mind before you reach eighteen."

The significant thing for us at present is that Lobachewsky changed the rules of Euclid's game and invented another just as good. This was a tremendous step forward.

It showed mathematicians that they might try the same trick of *denying the 'obvious,'* of ignoring or contradicting those things which have been accepted in any region of mathematics, "always, everywhere, and by all," and seeing what might come out of their boldness.

In geometry alone the outcome since Lobachewsky first showed the way was sufficiently staggering. Geometries by scores were created and studied intensively. When first made, these were created for their own sake. More than once these manufactured geometries proved invaluable in science, for which the classical geometry of Euclid was quite inadequate. I shall return to this later.

Before leaving this, however, let us notice another way in which geometry freed itself of the shackles of tradition by generalization. Solid space for the Greeks had *three* dimensions, say length, breadth, and thickness. When geometry began to be studied analytically or algebraically instead of synthetically, as had been the case up to 1637, the restriction to *three* dimensions no longer was necessary. It was only after the 1840s, however, that complete freedom was attained in this direction. First, in analytical mechanics in the eighteenth century, it became useful to reason about solid space and time together as a geometry of *four* dimensions. The step from *four* to *n* (*any* whole, positive, finite number) was taken by Cayley in 1843. The long stride from *n* dimensions to a *countable infinity* of dimensions was taken by Hilbert in 1906. I shall return to this in greater detail in subsequent chapters. A countable infinity is as many as there are of *all* the positive whole numbers, 1, 2, 3, . . . From geometry of a countable infinity of dimensions to an *uncountable infinity* (as many as there are points on a straight line) of dimensions, was the last step, taken about 1920.

If common sense objects to geometry of four dimensions,

it will get little comfort from modern physics. Relativity is based on a particular geometry of four dimensions, and geometry of an *infinity* of dimensions is now commonly used in the mechanics of atoms. All this will be resumed and discussed in more detail in Chapter 10.

The postulational method of setting up mathematical theories—algebra, geometry, and the rest—was one of the major additions to mathematics of the late nineteenth and early twentieth centuries, and it continues to be both useful and clarifying.

Chapter 4

"THE SAME, YET NOT THE SAME"

❖ *4.1.* ❖ *Realizations of Common Algebra*

No one with a musical ear would mistake a jig for a waltz.
The structure of each betrays its nature in the first few
bars or phrases. Nor would a musician confuse two waltzes.
Although they belong to the same kind of composition,
their melodies alone are sufficient to distinguish them
immediately.

In mathematics there is frequently discernible a similar
structure. Within each of several theories is an inner harmony of pure form, and the form for all is the same. But
two theories having the same abstract form may be as
different in their outward appearance and in their applications as are two waltzes in sound and emotional appeal.
This is not intended as more than a rough description, and
the analogy must not be pressed too far.

As a somewhat crude example, let us look first at the
postulates of a field stated in the preceding chapter.
We shall see that common algebra can be 'realized' in any
one of at least three ways. In the first the *class* concerned is
that of all *rational* numbers; in the second the class is that of
all *real* numbers; in the third the class is that of all *complex*
numbers. The *structure* of these three fields is the same,
namely Postulates (1.1), (1.2), (1.3), (2.1), (2.3), and
7. Each is, say, to pursue the analogy, a waltz; the tunes of
all three are different. If we work out the consequences of
the postulates once for all *abstractly*, without asking for a
tune to lighten our labors, we shall have done waltzes com-

pletely, all except fitting melodies to particular waltzes. The melodies correspond to the interpretations of the things in the given abstract class and those of the abstract operations according to which these things are combined in accordance with the postulates.

We use *abstract* to emphasize that *we can say nothing about the system considered,* here a field, *beyond what is explicitly stated in the postulates and what can be deduced by common logic from those postulates alone.* When we say, for example, that the things in the given classes are real numbers, we assert something which is not deducible from the postulates, for in them the things were mere marks. By thus putting a definite restriction on the marks, we get a field which is no longer *abstract* or *general,* but *special.* The formulas for this special field will be *instances* of those for the abstract field.

❖ 4.2 ❖ *Rational, Real, Denumerable, Non-denumerable, Discrete, Continuous, Complex, Analysis, Function*

This longest and most detailed heading in the book calls attention to several fundamental notions to which we shall frequently have to refer.

I must first describe what is meant by rational, real, and complex numbers. They permeate much of mathematics. It is assumed that we understand what the zero, positive, and negative whole numbers, $0, 1, 2, \ldots, -1, -2 \ldots$ are— a vast assumption in the light of modern critical mathematics [see Chapters 19, 20].

If a,b are whole numbers, of which b is not zero, the *ratio* of a to b is a/b (the result of dividing a by b). A *rational* number is defined as the ratio of two whole numbers. The class of all whole numbers is a subclass of all rational numbers, as is seen by restricting the divisor b to be 1.

The rational numbers do not include the irrationals. A number is called *irrational* if it is not the ratio of any pair of whole numbers. For example, the square root of 2, ($\sqrt{2}$), is irrational, as can be easily proved by supposing the contrary and getting a contradiction (for the proof see [19.4]). This fact, by the way, so disconcerted Pythagoras, who had constructed his theory of the universe on the hypothesis that all numbers are rational, that he induced its discoverer to drown himself in order to suppress the awkward theory-destroying fact. So runs the story. It is also reported that the fact had become so notorious in the golden age of Greece that Plato averred that anyone who did not know that $\sqrt{2}$ is irrational (he used different words, suited to geometry) was not a man but a beast.

A part of all the irrationals and all the rationals are swept up into the common class of *real* numbers. To picture these, take any convenient point O on an indefinitely extended straight line, and any convenient length, say an inch, which we agree shall be the unit of measure. Step off 1, 2, 3, . . . inches to the right of O, and 1, 2, 3, . . . to the left; name the first *positive*, and the second *negative*. The points thus marked, including 0 at O, correspond to the whole numbers. Scattered along the line are the points corresponding to the rational numbers, a few of which are marked in the figure. Where is $\sqrt{2}$ on the line? To the right of O and *somewhere between the two rational numbers* $\frac{140}{100}$ *and* $\frac{142}{100}$.

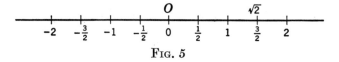

Fig. 5

Being content for the present with that vague 'somewhere,' we remark that *to each point on the line corresponds one and only one real number, rational or irrational.* The real numbers are everywhere dense on the line, for between any two we can always locate another—by bisecting the segment joining the two representative points, if no other way suggests itself. The class of all real numbers is the class whose members correspond, one-one, to all the points on the line. This intuitive notion of the real number system, imaged conceptually as just described, is of fundamental importance for an understanding of the numerous applications of mathematical analysis to science and technology. The class of all real numbers is called a *continuum* (of numbers)— "the Grand Continuum," as Sylvester dignified it in one of his expansive moments.

The infinitely numerous classes of the rational numbers and the real numbers introduced us to two different types of infinites. The infinity of all rational numbers can be enumerated, or counted off, in the order 1, 2, 3, . . . of the natural numbers, and so this infinity is said to be *denumerable* [3.3]. The infinity of all real numbers, which includes the infinity of all rational numbers, can *not* be counted off 1, 2, 3, . . . and is said to be *non-denumerable* [3.3].

It takes some ingenuity to prove these assertions, especially the second. I shall indicate only how the first goes. Imagine all the positive rational numbers $(1 = \frac{1}{1}, 2 = \frac{2}{1}, 3 = \frac{3}{1}$, and so on) $1, \frac{1}{2}, 2, \frac{1}{3}, 3, \frac{1}{4}, \frac{2}{3}, \frac{3}{2}, 4, \frac{1}{5}, \ldots$, where those numbers having a smaller sum of numerator and denominator precede those having a larger such sum. The sums in question for the above rationals are

$$2, 3, 3, 4, 4, 5, 5, 5, 5, 6, \ldots$$
$$(1, 2, 3, 4, 5, 6, 7, 8, 9, 10, \ldots)$$

where under the respective sums are written in parentheses the natural numbers 1, 2, 3, . . . corresponding to the ranks of the numbers in the sequence. Thus the seventh number is $\frac{2}{3}$, the tenth, $\frac{1}{5}$. Note that $\frac{4}{6}$, for example, does not occur in this form, because it is not reduced to its lowest terms, $\frac{2}{3}$, which does occur. It is clear that every rational number will appear only once in the sequence and will have a unique rank 1, or 2, or 3, ... So the rational numbers are denumerable. It is an interesting exercise in the elementary theory of numbers to find the rank of any given positive rational number, say $\frac{80}{231}$, and also what number has a given rank, say 1,000, without actually writing out the sequence.

Two further technical terms may be described here. A class or set [3.1] of things which can be counted off, 1, 2, 3, . . . , that is, which is denumerable, is said to be *discrete*. For example, the set of all the grains of sand on all the beaches in the world is discrete. A continuum of numbers, such as the set of all those corresponding to all the points on a line segment, is not a discrete set. A variable number which passes through the numbers in a set which is not discrete is said to be continuous. Although I realize that this needs elaboration, I trust that the essential meaning will appear as we proceed. More will be said about variables later [6.2], and numerous instances of what they actually are in scientific applications of mathematics will occur in subsequent chapters. Something, after all, must be left to language and sense even in a rigorous account of mathematics, which this sketch has no pretension to be.

Mathematical analysis, or briefly *analysis*, is concerned with *continuously* varying 'quantities' or numbers. For example, if x denotes a real number as described above, and if x can take values from a *continuum*, a mathematical expression involving x is a proper object for study in analysis. Thus $x, 1/x, x^2, . . . ,$ as x runs through all the numbers in the continuum from 0 to 1 (0 being excluded as a divisor) are investigated in analysis. To anticipate, expressions such

as those just described are called *functions* of the *variable x*.
Roughly, we say that the expression $f(x)$ (any formula involving x) is a function of x if $f(x)$ can be calculated when x
is assigned. Similarly for functions $f(x, y, \ldots)$ of variables $x, y, \ldots$.

Complex numbers constitute a still vaster assemblage. In
describing them, I shall deliberately avoid the perfectly
satisfactory way of the high-school texts and return to
Gauss. This method has two advantages for our purposes. It
avoids the legitimate but trivial discussion of what
'imaginary' means. 'Imaginary' numbers are no more
imaginary than are negatives, if we persist in regarding the
positive whole numbers as the only true numbers. It also
makes it easy to see how mathematicians beginning in the
1840s *generalized* complex numbers and invented numerous
systems of *hypercomplex numbers* [5.8].

Following Gauss, we let a,b represent any real numbers,
and we create an ordered *couple* (a, b). This ordered couple
of real numbers is called a *complex number* if it is made to
satisfy certain postulates, of which I shall state only three
as samples.

The *sum* $(a, b) \oplus (c, d)$ of the given pair of complex numbers (a, b), (c, d) is defined to be the complex number
$(a + c, b + d)$; the result $(a, b) \odot (c, d)$ of *multiplying* the
given pair (a, b), (c, d) is defined to be $(ac - bd, ad + bc)$;
'equality' is defined to mean that $(a, b) = (c, d)$ when and
only when $a = c$ *and* $b = d$. In the above, $a + c$, ac, and
so on, have their usual meanings as for real numbers in
arithmetic.

With these definitions of 'addition,' 'multiplication,' and
'equality,' it is a simple exercise to verify that the class of all
complex numbers (a, b), (c, d), $\ldots$ satisfies all the postulates of a field.

In passing, I give the usual geometrical picture of (a, b)

(Figure 6). Through O draw a perpendicular to the line on which we represented real numbers. Take any point P in the plane fixed by these two lines, and drop a perpendicular PN to the line of real numbers. If the length of ON is measured by the real number a, and that of NP when laid along the line of real numbers is measured by b, we *affix* to the point P the complex number (a, b). If P lies in either of the quadrants labeled I, IV, a is positive; if P lies in II, III, a is nega-

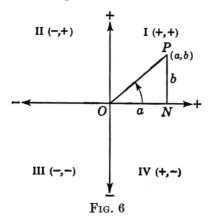

Fig. 6

tive; if P is in I or II, b is positive; if P is in III or IV, b is negative. The pairs $(+, +)$, $(-, +)$, $(-, -)$, $(+, -)$, taken in the order opposite to the motion of the hands of a watch, tell the story on the figure. The 'imaginary' $\sqrt{-1}$ has not been mentioned. Whoever cares to look for it will find its image on the vertical line. Notice that (a, b) is also uniquely placed by giving the length of OP and the magnitude of the angle NOP, read as indicated by the arrow. Now OP is a *vector*, whose *magnitude* is the length of OP and whose *direction* is NOP. This perhaps suggests why complex numbers are of great use in the study of alternating currents, where the vectors concerned are represented graphically.

In [5.3] I shall give yet another representation of com-

plex numbers and their algebra. Although this originated in
1847, it is more in the spirit of twentieth-century abstract
algebra than either of the classical representations described
above. It was long overlooked.

Out of all this several simple and important things
emerge. First, the infinitely rich class of all real numbers is
imaged on a mere straight line on the plane picturing the
class of all complex numbers, which is infinitely rich in
straight lines—they can be drawn in all directions over
the whole plane. To anticipate, I state here one of the great
discoveries of the late nineteenth century. Common sense
and all appearances notwithstanding, there are precisely as
many real numbers as there are complex numbers. Stated
geometrically, this says that on *any* straight line in a plane
there are just as many points as there are in the whole plane.
If that is not sufficiently jarring to the original sin of our
preconceived notions, consider this. In the whole plane
there are only as many points as there are on any segment
of a straight line, provided only that the segment is indeed
a segment and has a length not zero—say a billionth of a
billionth of a billionth of an inch. There is a still more striking conclusion of a similar sort. The segment contains as
many points as there are in the whole of space of a *countable*
(denumerable) infinity of dimensions.

If the reader will look back a few sentences he will see the
words 'one of the great discoveries of the late nineteenth
century.' That was not a mere rhetorical flourish. It was a
historical statement and was meant to be taken literally. It
was neither asserted nor implied that a great discovery is
ever necessarily the final one in a given direction. This discovery and what induced it led to a turning point in mathematics, and we do not yet know whether the signpost reads
'Go on' or 'Go back.' I shall return to this in Chapters 19
and 20.

❖ 4.3 ❖ *The End of a Road*

What else do the rational, the real, and the complex numbers give us, beyond a nest of Chinese boxes each of which is enclosed in the one following it? Every schoolboy knows or takes for granted that each of the first two classes of numbers satisfies all the demands of common algebra, and those slightly more advanced know the same for complex numbers. We have, then, three distinct instances of a field—three waltzes with different melodies. The *structure* of the field is the same in all three, which are *abstractly* identical; the *specialized* fields differ in their interpretations.

Before leaving this, I shall answer the natural question suggested by the rationals, the reals, and the complexes. Why not generalize still further, say to triples (*a, b, c*) of rationals, combined according to appropriate rules?

The answer is again a great landmark. It was proved by K. Weierstrass (1815–1897) about 1860, and more simply by Hilbert later, that *no further generalization in this particular direction is possible.* We have reached the end of a road. As it is of some importance to understand exactly what Weierstrass proved, I state it more fully. *By retaining ALL the postulates of a field, it is impossible to construct a class of things which satisfies all the postulates and which is not either the class of all complex numbers or one of the latter's subclasses.*

Here again I have tried to be historically precise. It *was* a landmark of the golden age. So broad, so rapid, and so deep was the river of mathematical progress, however, that this landmark was endangered. Not the fact which Weierstrass and Hilbert thought they proved was swept away, but the type of reasoning which they employed was called into serious question by the 1930s. Professing no opinion on these matters, which affect all our reasoning in logical patterns inherited from Aristotle [2.2], I simply report and pass on.

If complex numbers are the end of this particular road, how shall we progress? Go back and build another! New roads by hundreds were constructed to higher points of view by the mathematicians of the nineteenth and twentieth centuries. One great highway led to the unbounded field of *linear associative algebra*, in which the associative property of multiplication, but not the commutative, is retained in the postulates.

Having acquired from Lobachewsky, Hamilton, and others the habit of denying the 'obvious,' the pioneers might easily have contradicted or denied one or more of the postulates of a field, as we now sometimes do, to reach these vantage points. But this is rather a road for the sophisticated, easy enough to travel after it has been blasted out of the rock and graded. As a matter of fact one of the commanding peaks of the nineteenth century, which we could now reach more easily, was discovered otherwise and far from naturally. It all but revealed itself through the mists a score of times to seasoned explorers, who glimpsed its lower slopes but never its summit, for almost a century before a boy of eighteen looked up and saw it all. Less than three years later he was killed in a duel. From the summit which E. Galois (1811–1832) discovered, a host of workers, led by C. Jordan (1838–1922) and Kronecker, looked out over the vast domain of algebraic equations and algebraic numbers [11.5] and perceived order, simplicity, and beauty in what was chaos to the pioneers; another host, led by Klein and ascending yet higher, saw the geometries of his time united into a single coherent pattern of light and shade.

I shall indicate these summits in Chapter 8.

Chapter 5

THE ART OF ABSTRACTION

✤ 5.1 ✤ A Change of Interest

In the preceding chapter we saw that the field of complex numbers contains the fields of rational numbers and real numbers. Each of the last two is called a 'subfield' of the complex field. When we come to 'rings' and 'groups' we shall find 'subrings' and 'subgroups' contained in specified rings and groups. Such phenomena suggested in the 1900s to 1940s the study of a given algebraic 'variety'—such as a field, a ring, a group—in relation to its 'subvarieties'— such as subfields, subrings, subgroups.

So far—in fields, for instance—our interest has been in the elements a, b, c, . . . of the entire variety, rather than in the relations between the relevant subvarieties considered as elements of a new variety. An extensive department of modern algebra is concerned with the relations between subvarieties. It began to be intensively developed for its own interest in the 1930s, and shortly passed beyond algebra, clarifying the rudiments of some other departments of mathematics. It was implicit as early as 1831 in the work of Galois, and almost explicit in the algebra of logic initiated in 1847–1854 by G. Boole (1815–1864). These things among others, including projective geometry, of nineteenth-century mathematics finally evolved into a theory of 'structures' or 'lattices.'

To see how this came about we must go into some detail for specific examples and special cases of a few of the concepts ultimately abstracted in the algebra of structures.

These and many others showed that common to several departments of algebra there are certain extremely simple underlying laws of combination abstractly the same for all. The art of abstraction is to discover and isolate these laws. When discovered, the laws are expressed in a symbolism that can be variously specialized, or interpreted, so that at least the leading notions of each of the departments of algebra from which the laws were abstracted is recovered. Likewise for divisions of mathematics other than algebra. The material available for illustration is so extensive that a selection must be made. I shall choose items that seem to promise some lasting interest.

As I have alluded in passing to mathematical or symbolic logic, and as this was and is one of the plainest clues to an algebraic theory of structure, I shall describe it first. In some respects it is like common algebra, in others strikingly different.

❖ 5.2 ❖ *An Uncommon Algebra*

The philosopher, mathematician, and diplomat Leibniz, in what he called a "schoolboy essay," written when he was sixteen and published in 1666, imagined the possibility of a "universal characteristic," or "kind of universal mathematics" to guide the very fallible human reason. Purely verbal arguments in mathematics are usually harder to follow than those employing some of the customary symbolism. Leibniz hoped to reduce all errors of reasoning to trivial slips in calculation. Although he made some progress toward a symbolic reasoning and had a few followers who went a little farther, the first really significant advance came only in 1854, when Boole published *An investigation of the laws of thought.* This in its turn attracted but little attention from the busy and prolific mathematicians of the nineteenth century. However, the reduction of certain types of logical and mathematical reasoning to symbolism continued to gather speed and by 1910, when the first volume of *Principia mathematica* by A. N. Whitehead (1861–1947) and Russell appeared, symbolic logic and its applications to the foundations of mathematics were well-established disciplines of technical mathematics itself. Leibniz' project of reducing *all* reasoning to a kind of mechanical mathematics was replaced by the much less ambitious but still sufficiently difficult task of eliminating subtle mistakes in the simplest traditional mathematics, for example, common arithmetic.

It will be enough here to indicate how it is possible to reduce at least the rudiments of logical reasoning to a species of algebra having certain of its formal laws in common with some of those of a field, in particular with the associative and commutative laws of addition and multiplication [3.2]. There are numerous sets of postulates for the elementary algebra of logic, and almost as many systems of notation. That chosen in what immediately follows is Huntington's first set (*Transactions of the American Mathematical Society*, Vol. 5, pp. 292–293, 1904). (∧, ∨ as here are not to be confused with the same in other accounts, such as, among many more, A. Tarski's in his readable *Introduction to Logic*, 2nd edition, where ∧, ∨ are ⊙, ⊕ here.) Beginning at the historical tail instead of the head, we shall consider the following ten postulates (1.1) to (6) which evolved from the pioneering work of the nineteenth century.

The fundamental concepts are a *class* or *set* K and *two rules of combination* ⊕, ⊙. The things in K are the *elements* of K.

(1.1) $a \oplus b$ is in the class whenever a,b are in the class.

(1.2) $a \odot b$ is in the class whenever a,b are in the class.

The equality sign, $=$, in what follows is assumed to have the following properties. If a,b are in K, exactly one of

$a = b$ ('a equal to b') $a \neq b$ ('a not equal to b') holds; $a = a$; if $a = b$, then $b = a$; if a,b,c are in K, and if $a = b$ and $b = c$, then $a = c$. If $a,b,c, \ldots$ are subsets of a class K, an interpretation of $a = b$ is, 'everything that is in a is in b, and everything that is in b is in a,' or 'a includes b' and 'b includes a.'

(2.1) There is an element $\wedge$ such that $a \oplus \wedge = a$ for every element a.

(2.2) There is an element $\vee$ such that $a \odot \vee = a$ for every element a.

(3.1) $a \oplus b = b \oplus a$ whenever a, b, $a \oplus b$, and $b \oplus a$ are in the class.

(3.2) $a \odot b = b \odot a$ whenever a, b, $a \odot b$, and $b \odot a$ are in the class.

(4.1) $a \oplus (b \odot c) = (a \oplus b) \odot (a \oplus c)$ whenever a, b, c, $a \oplus b$, $a \oplus c$, $b \odot c$, $a \oplus (b \odot c)$, and $(a \oplus b) \odot (a \oplus c)$ are in the class.

(4.2) $a \odot (b \oplus c) = (a \odot b) \oplus (a \odot c)$ whenever a, b, c, $a \odot b$, $a \odot c$, $b \oplus c$, $a \odot (b \oplus c)$, and $(a \odot b) \oplus (a \odot c)$ are in the class.

(5) If the elements $\wedge$, $\vee$ in (2.1), (2.2) exist and are unique, then for every element a there is an element $\bar{a}$ such that $a \oplus \bar{a} = \vee$ and $a \odot \bar{a} = \wedge$.

(6) There are at least two elements, x and y, in the class such that $x \neq y$ (x is not 'equal to' y).

From these with patience and some ingenuity any desired number of theorems can be deduced. Among the simpler and more important for the interpretations in logic are these.

(7.1) The element $\wedge$ in (2.1) is unique: $a \oplus \wedge = a$.

(7.2) The element $\vee$ in (2.2) is unique: $a \odot \vee = a$.

(8.1) $a \oplus a = a$.

(8.2) $a \odot a = a$.

(9.1) $a \oplus \vee = \vee$.

(9.2) $a \odot \wedge = \wedge$.
(10.1) $a \oplus (a \odot b) = a$.
(10.2) $a \odot (a \oplus b) = a$.
(11) The element $\bar{a}$ in (5) is uniquely determined by a,

$$a \odot \bar{a} = \vee, \, a \odot \bar{a} = \wedge.$$

This element $\bar{a}$ is called the *complement* of a. It follows readily that the complement $\overline{(\bar{a})}$ of the complement $\bar{a}$ of a is a. Complementation is thus of *period* 2, like subtraction.

(12.1) $a \oplus b = \overline{\bar{a} \odot \bar{b}}$.
(12.2) $a \odot b = \overline{\bar{a} \oplus \bar{b}}$.
(13.1) $(a \oplus b) \oplus c = a \oplus (b \oplus c)$.
(13.2) $(a \odot b) \odot c = a \odot (b \odot c)$.

The similarities between some of these postulates and theorems and some of those for a field—for example, (3.1), (3.2), (4.2), (13.1), (13.2)—are plain. The symmetrical dualities in the pairs of statements, with the accompanying interchanges of $\wedge$, $\vee$ where relevant, also are evident. The pair (12.1), (12.2) gives the general *principle of duality* between $\oplus$, $\odot$. The commutative laws for $\oplus$, $\odot$ are postulated in (3.1), (3.2); a distributive law is postulated in (4.2), and another, peculiar to this algebra as distinguished from a field, in (4.1); while (2.1), (2.2) are as in a field when $\wedge$, $\vee$ are interpreted as 0, 1. *The associative laws* (13.1), (13.2) *are not postulates here but theorems.* The striking divergences between this algebra and a field issue largely from the theorems (8.1), (8.2). Replacing $\oplus$ by $+$ and $a \odot b$ by ab for convenience, we see that these are $a + a = a$, $aa = a$ or, say, $2a = a$, $a^2 = a$ for every a in the class. So in this *boolean algebra of two elements*, 0, 1, the coefficients and exponents in polynomials are limited to the numbers 0, 1, with $0a = \wedge = 0$, $a^0 = \vee = 1$. Theorem (10.1), or

$$a(a + b) = a,$$

that is, $a + ab = a$, is called the *law of absorption;* why, will be clear when we state a historical and one current interpretation of this algebra.

We have just seen that any element x in this algebra is idempotent, $x^2 = x$. This is characteristic of a commutative ring with unit element (defined in the next section) which is a *boolean algebra.*

Again it is easily proved that in our simplified notation

$$a\bar{a} = 0, a + \bar{a} = 1.$$

I note these particularly because verbal logical equivalents of them suggested Boole's algebra to him as early as 1847. The first, as will be evident presently, is Aristotle's logical law of the excluded middle, the second his law of contradiction as in classical logic.

What is to be made of this uncommon algebra? The consistency of the ten postulates on which it is based is fairly easily proved *without reference to the historical origin of the postulates.* To be optimistic (or pessimistic?) for a moment, let us suppose that the human race is not extinct five hundred years hence and that by some oversight a copy of these postulates has survived the latest mass insanity. If history is any guide, arithmetic and algebra will be reinvented long before symbolic logic evolves again, if it ever does. What will the baffled algebraists make of the ten postulates? As a charitable guess, the saner among them will dismiss the postulates as the work of a disordered mind or the garbled notes of a backward student scribbled down from what the football coach dictated when the teacher of algebra was on sick leave. We, with history behind instead of ahead of us, of course know better.

It is readily verified that this algebra makes sense when the elements $a, b, \ldots$ of the class K are themselves classes and $\wedge$ is the null class (the class having no members), $\vee$ is the class of all subsets (or subclasses) of K, including $\wedge$ and K itself; $a \oplus b$ is the class of all things that are in a or in b, or in both a and b, and $a \odot b$ is the class of all things that are in both a and b, and $\bar{a}$ is the class of all things that are not in a. With this interpretation (8.1) — (9.2), for example, are truisms.

A second interpretation, derivable from the first, concerns *propositions*. It is sufficient here to think of a proposition as a statement which is a definite and ascertainable one of 'true' (T), 'false' (F), although we do not go into the metaphysics of what 'true,' 'false' may signify, if anything, in 'reality.' 'True' in mathematics is sometimes identified with consistency or freedom from contradiction.

Instead of deriving the second interpretation from the first I shall make a fresh start, abandoning $a, b, c, \ldots$ for propositions $p, q, r, \ldots$ and formulating a selection of the basic notions of the logic, or calculus, of propositions in one of many current symbolisms for 'not,' 'and,' 'or,' 'if-then,' the last being 'implication.' The entire system can be regarded abstractly as a pure formalism. However, having in mind the interpretation in terms of a *finite number* of propositions will make the definitions seem less arbitrary and the theorems less arid than they otherwise might. For example, 'p and q' in the interpretation is the joint assertion of p,q; 'not-p' is the denial of p. If p is true, 'not-p' is false; if p is false, 'not-p' is true. Perhaps the simplest and clearest way of stating the definitions is by means of 'truth tables,' or 'T,F tables,' where T in the interpretation considered here is 'true' and F is 'false.' The device of T,F tables, was invented by C. S. (S.) Peirce (1839–1914) but was overlooked or forgotten until the 1920s, when it was reinvented. An advantage of this device is that it immediately suggests a generalization: 'truth values' of proposi-

tions are not confined to the *two T,F*. This will be described
after we have looked at the two-valued system.

The *p, q*, . . . are variables (propositional variables),
and in a statement such as '*If p, then p or q*' the subsidiary
statement, or 'universal quantification,' '*for any p, q*' is to
be understood. This applies in particular to the truth
tables given immediately.

For ease in printing 'not-*p*' is written p' and is defined
by the table

$$\begin{array}{c|c} p & p' \\ \hline T & F \\ F & T \end{array}$$

In the interpretation this states that if p is true, 'not-p' is
false; if p is false, 'not-p' is true.

The definitions of '*p and q*,' symbolized $p \cap q$; '*p or q*,'
symbolized $p \cup q$; '*p implies q*,' symbolized $p \Rightarrow q$; and
'*p is equivalent to q*,' symbolized $p \Leftrightarrow q$, are read off from
the table

p	q	$p \cap q$	$p \cup q$	$p \Rightarrow q$	$p \Leftrightarrow q$
T	T	T	T	T	T
F	T	F	T	T	F
T	F	F	T	F	F
F	F	F	F	T	T

The only one of these that seems when interpreted to
diverge from traditional common sense is $p \Rightarrow q$, which can
be read 'if p, then q.' The column for $p \Rightarrow q$ when interpreted
says that p implies q if p is false or q is true, and incidentally
that a false proposition implies any proposition. For example, 'the moon is made of green cheese' implies that 'twice
two is four.' But 'twice two is four' does not imply that
'the moon is made of green cheese.' Some of the classical

logicians objected to this definition of implication on metaphysical grounds, overlooking the fact that it is futile to
quarrel with a definition. Implication was defined precisely and formally as above because the definition was
found expedient in actual work in mathematical logic. It
does not contradict tradition but supplements it. The definition of $p \Leftrightarrow q$ shows that it is the same as p *implies* q,
and q *implies* p. If this is not immediately evident we proceed as follows, using the definitions of the preceding table,
and so in all similar instances.

p	q	$p \Rightarrow q$	$q \Rightarrow p$	$(p \Rightarrow q) \cap (q \Rightarrow p)$
T	T	T	T	T
F	T	T	F	F
T	F	F	T	F
F	F	T	T	T

The columns headed p,q exhaust the T,F, possibilities for
2 propositions p,q. If instead of 2 propositions there are 3,
p,q,r, the table has 8 ($= 2^3$) rows; for 4 propositions the
table has 16 ($= 2^4$) rows, and so on. The T or F of any compound proposition constructed from p', $p \cup q$, $p \cap q$, and
so on, as defined above, *can be ascertained by purely mechanical operations* as in the above example. A proposition which
is T for all the possible T,F choices of the propositions p,
$q, r, \ldots$ from which it is constructed is called a *tautology*
and is said to be a *law* of the system; one which is F for
all choices is called a *contradiction*. Tautologies are more
highly prized than contradictions.

Purely formal algebras or calculating machines can be,
and have been, described or constructed for grinding out
tautologies like sausages indefinitely. Machinery enters with
T,F, a *two*-valued logic and thence, among other ways, with
an open or closed circuit. The laws of contradiction and
excluded middle assure us that a two-way switch must be

open or closed but not both. These laws of classical logic are the brain of these modern calculating machines called 'digital'—to be noted in [11.8]. Conversely, this logical algebra was applied (1937–) to the design of complicated electrical circuits.

From the mere mechanics of the situation, a generalization to n-way switches or electronic tubes and n-valued 'truth systems' is suggested. For $n = 3$, the 'truth values' may be labeled T ('true'), F ('false'), and D ('doubtful'). 'Implication,' 'and,' 'or,' 'not,' and so on are defined in this system as indicated by various kinds of problems. One in particular proved of some scientific interest. Heisenberg's uncertainty principle of 1927 in quantum mechanics (see [18.4]), suggested a three-valued logic. This was developed (1944) by H. Reichenbach (1891–). Although no new physics was discovered thus, it appeared that it was futile to search in certain directions for coherent theories to correlate experimental facts. The three-valued logic is said to have been imagined by the logicians and theologians of the Middle Ages, but was first proposed in modern times in 1920–21 by Tarski, J. Lucasiewicz (contemporary), and E. L. Post (contemporary). Before long, four-, five-, . . . valued logics appeared by more or less immediate algebraic generalization of the truth tables for the case $n = 2$, and from these followed the inevitable generalization to logics with either a denumerable or a non-denumerable infinity [4.2] of truth values. A remarkable algebraic feature of the case $n = 2$ carries over to any n-valued system, n or integer: all the basic notions, such as 'not,' 'and,' 'or,' and so on, can be generated by iterations of a single operation; the entire calculus of propositions in an n-valued logic can be generated by this operation.

It will be interesting to see the correlates for propositions $p, q, r, \ldots$ of some of the postulates and theorems (1.1)–

(13.2) as they were interpreted for classes a, b, c, ... Equality, $=$, is replaced by $\Leftrightarrow$, $\oplus$ by $\cup$, $\odot$ by $\cap$; $\wedge$, $\vee$ and $\bar{a}$ may be left to the ingenuity of the reader. For (3.1), (3.2) the correlates are the tautologies $p \cup q \Leftrightarrow q \cup p$, $p \cap q \Leftrightarrow q \cap p$. The correlates of (8.1), (8.2) are $p \cup p \Leftrightarrow p$, $p \cap p \Leftrightarrow p$. The first of the latter states that '*If p or p then p, and, if p then p and p.*' This can be verified by the method of truth tables. A somewhat livelier example is furnished by (10.2)

$$p \cup (p \cap q) \Leftrightarrow p:$$

p	q	$p \cap q$	$p \cup (p \cap q)$	$p \cup (p \cap q) \Leftrightarrow p$
T	T	T	T	T
F	T	F	F	T
T	F	F	T	T
F	F	F	F	T

So this proposition is a tautology. To see that it is not entirely trivial the reader may turn it into English and prove it verbally. Likewise for our final example, involving three propositions p,q,r,

$$[p \cap q \Rightarrow r] \Rightarrow [p \Rightarrow (q' \cup r)].$$

To make the following table printable without a break, I shall call this compound proposition A.

p	q	r	$p \cap q$	$p \cap q \Rightarrow r$	$q' \cup r$	$p \Rightarrow (q' \cup r)$	A
T	T	T	T	T	T	T	T
T	T	F	T	F	F	F	T
T	F	T	F	T	T	T	T
T	F	F	F	T	T	T	T
F	T	T	F	T	T	T	T
F	T	F	F	T	F	T	T
F	F	T	F	T	T	T	T
F	F	F	F	T	T	T	T

In a sense that needs no elaboration each of p' (not-p), $p \cap q$ (p and q), and so on is a function of p,q. This suggests a calculus of propositional functions, in which the variables are propositions p, q, . . . and the functions are propositions. If $f(p)$ is a proposition whenever p is a proposition, $f(p)$ is called a *propositional function;* for example, $f(p) = p'$. Similarly for $f(p, q, . . .)$, where p, q, . . . are propositions; for example, $f(p, q) = p \cup q$. In this calculus two notions are basic, 'for all,' and 'there exists.' Thus if for all p and for all q, $f(q, p)$ is such that $f(p, q) \Leftrightarrow f(q, p)$, f is said to be symmetric, and we write

$$(p)(q)[f(p, q) \Leftrightarrow f(q, p)],$$

where $(p)(q)$ is to be read 'for all p, for all q.' The (p), (q) are called *quantifiers*. The frequently occurring mathematical phrase 'there exists' is symbolized as Ⴈ. For example, if $f(x, y)$ is defined for x,y, and

$$(x)[(\textrm{Ⴈ}y)f(x, y) \cup (\textrm{Ⴈ}y)f(y, x) \Leftrightarrow f(x, x)],$$

it is seen that $f(x, y)$, defines the general *reflexive relation* R,xRx. An example of such a relation is equality, where R is $=$.

From here on the development becomes too technical for brief description. What is the purpose of it all, and what has it achieved? The purpose has been to capture and fix common though elusive notions of habitual logical and mathematical reasoning. The achievement is a sharper understanding of all logical reasoning, including mathematics, and a realization of what can or cannot be proved within a prescribed mathematical or logical context. This will be noted in Chapter 20; without the algebra of logic, the outstanding achievements of modern mathematics noted there would probably have been undiscovered.

❖ 5.3 ❖ Rings

We observed that one way of generalizing a mathematical system is to weaken the postulates by dropping one or more. The system generated by the weakened postulates is consistent if the original is, and it is a generalization of the original system. A generalization may be latent for many years in a familiar system without attracting much attention until new problems demand that it be studied intensively.

Such was the case with rings, which are generalizations of a field, and with 'lattices' or 'structures,' to be described later. A simple example of a (commutative) ring (with a unity element) is furnished by the infinite set of numbers . . . , -3, -2, -1, 0, 1, 2, 3, . . . and their properties with respect to the *two* 'operations' *addition* and *multiplication*. Subtraction is included; division is not. This set is closed under addition and multiplication, $a + b$ and ab are in the set whenever a,b are, and $a + b = b + a$, $ab = ba$. Also there is a unique number, namely 0, in the set such that $a + 0 = a$ for all a in the set, and for each number x in the set there is a unique number, denoted by $\bar{x}$, or $-x$, the *negative* of x, in the set, such that $x + \bar{x} = 0$. For multiplication, there is a unique number, 1, in the set such that $1a = a$ for all a in the set. In the general definition of a ring neither the commutativity of multiplication, $ab = ba$, nor the existence of a 1 ('unity element,' or 'identity element') for multiplication is assumed. The reason for this is that there are numerous mathematically interesting or scientifically important instances of more general rings; for example, matrices [6.3] under appropriate rules of combination. The formal definition follows.

A *ring* is a set S of elements a, b, c, . . . , z, . . . which can be combined according to two operations, 'addition,' $+$, and 'multiplication,' $\times$ ($a \times b$ is written ab), to yield uniquely determined elements of S, satisfying the following postulates.

(1.1) $a + b = b + a$ for all a,b in S.

(1.2) There is a unique element z in S such that, for each a in S, $a + z = a$. (z is called the 'zero' of the ring. It may be written 0 when there is no danger of confusion.)

(1.3) For each a in S, there is a unique element $-a$ in S, the 'negative' of a, such that $a + -a = z$. This is written more concisely as $a - a = 0$.

(1.4) For all a,b,c in S, $a(b + c) = ab + ac$,

$$(b + c)a = ba + ca.$$

These are the distributive laws of multiplication with respect to addition. Both are postulated, as commutativity of multiplication is not assumed.

If in addition to (1.1)–(1.4) it is postulated that for all a,b in S, $ab = ba$, the ring is called *commutative*. If there is an element e in S such that $ea = ae = a$ for all a in S, the ring is called a *ring with unit, or unity, element e*. We have given one example of a commutative ring with a unity element. Another, from a less specialized ring of importance in modern algebra and in physics, will be given when we come to matrices [6.3]. For the moment we note a peculiarity of some rings that does not occur in fields.

If $ab = z$ [z as in (1.2)] while $b \neq z$, a is called a *divisor of zero*, and is said to be a *proper* divisor of zero if $a \neq z$. To see that these definitions are not vacuous we may look at an example, actually the historically first. This example when elaborated suggested many ideas of modern algebra, including that of a general *equivalence relation* which I describe first. The essential idea goes back to Gauss in his masterpiece, the *Disquisitiones arithmeticae*, published in 1801 when Gauss was twenty-four.

A relation, written $\sim$, is said to be *binary* with respect to the elements a, b, c, . . . , x, . . . of a given class K of things if $a \sim b$ ('a is equivalent to b') is an ascertainable one of 'true,' 'false' for all a,b in the class. The binary relation $\sim$ is called an equivalence relation if the postulates (R), (S), (T) are satisfied.

(R) $a \sim a$ for every a in K. (Reflexivity.)

(S) If $a \sim b$, then $b \sim a$ for all a,b in K. (Symmetry.)

(T) If $a \sim b$ and $b \sim c$, then $a \sim c$. (Transitivity.)

The simplest example of an equivalence relation is equality, for which we have seen an example in [5.2]. An equivalence relation $\sim$ for the class K separates all the elements of K into mutually exclusive classes, called *equivalence classes*, C_a, C_b, . . . , C_x, . . . , all those elements of K that are equivalent to x, and only those, constituting C_x.

A less banal example than equality of an equivalence relation is that of *congruence* for the numbers . . . , -3, -2, -1, 0, 1, 2, 3, . . . This suggested several ideas of modern algebra, some of which will appear as we proceed.

Two numbers a,b are said to be *congruent* with respect to the (fixed, different from zero) number m, called the *modulus*, if the difference of a,b is divisible by m, and this is written

$$a \equiv b \bmod m,$$

which may be read 'a is congruent to b modulo m.' For example, $100 \equiv 0 \bmod 2$; $100 \equiv 1 \bmod 11$; $25 \equiv 8 \bmod 17$; $38 \equiv 26 \bmod 12$; $-100 \equiv 10 \equiv -1 \bmod 11$; $-25 \equiv 9 \bmod 17$. It is easily verified that $\equiv$ is an instance of $\sim$ in (R), (S), (T); that is, congruence is an equivalence relation. Further, for the modulus m there are precisely the m equivalence classes C_0, C_1, . . . , C_{m-1}, where C_a, say, contains all those numbers, and only those, that are congruent to a modulo m. Note that a in C_a is non-negative. It is an immediate consequence of the definition that congruence is preserved under addition, subtraction, and multiplication: if $x \equiv a \mod m$ and $y \equiv b \mod m$, then $x \pm y \equiv a \pm b \mod m$, and $xy \equiv ab \mod m$. From this we are led to define an *addition*, $C_a \oplus C_b$, and a multiplication, $C_a \odot C_b$, for the m equivalence classes just defined: $C_a \oplus C_b$ is the unique equivalence class C_r, and $C_a \odot C_b$ is the unique equivalence class C_s, where $a + b \equiv r \mod m$, and $ab \equiv s \mod m$, where the non-negative numbers r, s do not exceed m. Two classes are *equal* if and only if the numbers in both are the same.

To put some flesh on all these postulational bones, let us look at the cases $m = 5$, $m = 6$, noting that 5 is a *prime*, and 6 a *composite* number. For $m = 5$ the equivalence classes are C_0, C_1, C_2, C_3, C_4; $C_0 \oplus C_1 = C_1$, $C_2 \oplus C_4 = C_1$, $C_4 \oplus C_4 = C_3$, and so on. The addition and multiplication tables are

$\oplus$	C_0	C_1	C_2	C_3	C_4
C_0	C_0	C_1	C_2	C_3	C_4
C_1	C_1	C_2	C_3	C_4	C_0
C_2	C_2	C_3	C_4	C_0	C_1
C_3	C_3	C_4	C_0	C_1	C_2
C_4	C_4	C_0	C_1	C_2	C_3

$\odot$	C_0	C_1	C_2	C_3	C_4
C_0	C_0	C_0	C_0	C_0	C_0
C_1	C_0	C_1	C_2	C_3	C_4
C_2	C_0	C_2	C_4	C_1	C_3
C_3	C_0	C_3	C_1	C_4	C_2
C_4	C_0	C_4	C_3	C_2	C_1

From these, or from the definition of congruence, it follows readily that addition and multiplication in this example are commutative and associative, and multiplication

$\odot$ is distributive with respect to addition $\oplus$. Moreover C_0 is the unique zero element for addition, and C_1 the unique unity element for multiplication; and for any C_a there is a unique C_x such that $C_a \oplus C_x = C_0$. From the multiplication table, for any C_a, other than C_0 there is a unique C_y such that $C_a \odot C_y = C_1$. All this proves that for $m = 5$ the equivalence classes are a field with respect to $\oplus$, $\odot$. This field is an example of a *finite* field, one containing only a finite number of elements. The general theory of finite fields was worked out in the nineteenth and early twentieth centuries.

For $m = 6$, the tables are

$\oplus$	C_0	C_1	C_2	C_3	C_4	C_5
C_0	C_0	C_1	C_2	C_3	C_4	C_5
C_1	C_1	C_2	C_3	C_4	C_5	C_0
C_2	C_2	C_3	C_4	C_5	C_0	C_1
C_3	C_3	C_4	C_5	C_0	C_1	C_2
C_4	C_4	C_5	C_0	C_1	C_2	C_3
C_5	C_5	C_0	C_1	C_2	C_3	C_4

$\odot$	C_0	C_1	C_2	C_3	C_4	C_5
C_0	C_0	C_0	C_0	C_0	C_0	C_0
C_1	C_0	C_1	C_2	C_3	C_4	C_5
C_2	C_0	C_2	C_4	C_0	C_2	C_4
C_3	C_0	C_3	C_0	C_3	C_0	C_3
C_4	C_0	C_4	C_2	C_0	C_4	C_2
C_5	C_0	C_5	C_4	C_3	C_2	C_1

As in the preceding example it can be verified that these, the equivalence classes for $m = 6$, are a commutative ring with respect to $\oplus$, $\odot$; the ring has the unity element C_1 for $\odot$. This ring is not a field, as shown by the multiplication table; C_2, C_3, C_4 have no inverses. If they had, C_1 would occur in the corresponding rows of the table. This ring has proper divisors of zero, C_2, C_3, C_4:

$$C_2 \odot C_3 = C_0, \ C_3 \odot C_4 = C_0.$$

These two examples illustrate a general theorem. The equivalence classes for the modulus m are usually called *congruence classes* modulo m. It can be shown without much trouble that the congruence classes modulo m are a field with respect to $\oplus$, $\odot$ if, and only if, m is prime. If m is composite, the classes are a commutative ring having divisors of zero.

A commutative ring without divisors of zero is called a *domain of integrity,* or an *integral domain.* A simple example is the set of numbers 0, ± 1, ± 2, ± 3, $\cdot\cdot\cdot$.

An interesting example of congruence classes is the algebra of complex nunbers when mapped onto the residue classes modulo $x^2 + 1$ of the set of all polynomials in x with coefficients in the field of real numbers. The residues are of the form $a + bx$, a,b real numbers; the set of all polynomials congruent mod $x^2 + 1$ to $a + bx$ is denoted by $\{a + bx\}$; addition and multiplication are defined by

$$\{a + bx\} + \{c + dx\} = \{(a + b) + (c + d)x\};$$
$$\{a + bx\}\{c + dx\} = \{r + sx\},$$

where $ac + (ad + bc)x + bdx^2 \equiv r + sx \mod x^2 + 1$, and hence, since $x^2 \equiv -1 \mod x^2 + 1$,

$$r = ac - bd, \ s = ad + bc.$$

This example was worked out in detail by A. L. Cauchy (1789–1857) in 1847 to get rid of the sign $\sqrt{-1}$, to which he objected because "*it has no sense.*" Cauchy's objection did not register till about forty years after he was dead, when Kronecker and other algebraists expanded it into a theory of algebraic equations and algebraic numbers. It has yet to register in elementary texts.

❖ 5.4 ❖ *Homomorphism, Isomorphism, Automorphism*

These three notions are of frequent occurrence in modern algebra and in other divisions of mathematics. As the etymology of the words indicates, 'homomorphism' is similarity of form, 'isomorphism' is identity of form, while 'automorphism' is an isomorphism of a system to itself.

Let V, V' be two algebraic varieties, such as rings, for instance, in each of which operations having the formal properties of 'addition' and 'multiplication' as in a ring are defined, say $\oplus$, $\odot$ for V and $\oplus'$, $\odot'$ for V'. The basic notion is that of a *correspondence* between the elements of V and those of V' *which preserves sums and products*. The customary symbol for correspondence is $\rightarrow$; $a \rightarrow b$ is read 'a corresponds to b.'

The varieties V, V' are said to be *homomorphic* with respect to two operations, 'addition' and 'multiplication,' if there is a correspondence $a \rightarrow a'$ between the elements a, b, c, . . . of V and the elements a', b', c', . . . of V' such that each element a in V has *a unique* correspondent a' in V', and each element a' of V' has *at least one* correspondent in V, and the correspondence is such that when $a \rightarrow a'$ and $b \rightarrow b'$, then

$$a \oplus b \rightarrow a' \oplus' b', \; a \odot b \rightarrow a' \odot' b'.$$

The correspondence *maps* V on V'; a' is the *image* in V' of a in V. A simple example is given by V the ring of integers, V' the ring of congruence classes modulo m. All the numbers congruent to the same number map onto the same congruence class; the addition and multiplication in the map are as already defined for congruence classes.

If the correspondence is one-one, written $a \leftrightarrow a'$, so that $a \rightarrow a'$ and $a' \rightarrow a$, and hence each element of V maps into one element of V', and each element of V' maps into one element of V, the homomorphism is called an *isomorphism,* addition and multiplication being preserved as before.

As an illustration of the ideas described so far, I state the following theorem, leaving the simple proof to the ingenuity of the reader. If R and S' are two rings having no element in common, and if S' contains a subring R' isomorphic to R, then there is a ring S isomorphic to S' which contains R as a subring.

An *automorphism* is an isomorphism which maps a variety V on itself, that is, a one-one correspondence between the elements a, b, c, d, . . . of V such that if $a \leftrightarrow b$ and $c \leftrightarrow d$, then $a + c \leftrightarrow b + d$ and $ab \leftrightarrow cd$.

Without going into much detail, I shall give an example of automorphism for a field. Let m be an integer with all different prime factors, say $105(= 3 \times 5 \times 7)$. It is easily shown that the set of all numbers of the form $a + b \sqrt{m}$, where a,b are rational numbers is a field, the possibility $a = 0$, $b = 0$ being excluded in division. The *identity* (or *identical*) *automorphism* of this field is the trivial one

$$a + b \sqrt{m} \leftrightarrow a + b \sqrt{m}.$$

The only other automorphism is

$$a + b \sqrt{m} \leftrightarrow a - b \sqrt{m}.$$

To check that this is an automorphism with respect to addition, we observe that

$$(a + b \sqrt{m}) + (c + d \sqrt{m}) = (a + c) + (b + d) \sqrt{m},$$
$$(a - b \sqrt{m}) + (c - d \sqrt{m}) = (a + c) - (b + d) \sqrt{m};$$

and with respect to multiplication,

$$(a + b \sqrt{m})(c + d \sqrt{m}) = ac + bdm + (ad + bc) \sqrt{m},$$
$$(a - b \sqrt{m})(c - d \sqrt{m}) = ac + bdm - (ad + bc) \sqrt{m},$$

showing that correspondents of sums (or of products) map into sums (or products) of correspondents under the automorphism.

Similar definitions of isomorphism and automorphism hold for varieties in which only one operation is defined; for example (as will be seen [9.2]), for groups. For those already acquainted with groups, it is an easy exercise to prove that the automorphisms of a group are themselves a group, the *automorphism group* of the given group, when the product of two automorphisms is defined as the result, an automorphism, of applying the two successively. This is mentioned in passing because it is basic in the abstract reworking of the Galois theory (1831) of algebraic equations [9.8]. I shall return to isomorphisms and automorphisms after I have described groups, and will give further examples [9.2].

❖ 5.5 ❖ *Lattices or Structures*

The preceding examples were selected from a great many having similar characteristics in some of their details, partly for their own interest and partly to prepare the way for one of the more successful unifying notions of modern abstract algebra—lattices.

I remarked that abstract algebra quite suddenly became of interest in the 1920s. Actually some of it was implicit in J. W. R. Dedekind's (1831–1916) 'dual groups' of 1900, which need not be described here—an illustration of what was said about generalizations remaining latent for many years until new demands force them out. The need for a rigorous abstract algebra was acute in algebraic geometry long before it became available in the 1930s and 1940s for application to the intricate geometry of curves and surfaces as those vast theories had previously been developed. More than a hint of one active department of this abstract algebra—specifically, the theory of lattices or structures—was implicit in Boole's algebra of classes (1854) as modified and developed by his successors into what is now called boolean algebra. Analogies between boolean algebra and properties of the G.C.D. (greatest common divisor) and L.C.M. (least common multiple) as in elementary arithmetic were noted by the mathematical logicians as early as 1912. But these and the corresponding and more suggestive properties of algebraic numbers [11.5] were not followed up, and it was only in the mid-1930s that the theory of lattices really got under way, mainly in the work of G. Birkhoff (1911–) and O. Ore (1899–). The theory has revealed many abstractly identical details of theories which at first sight may seem to have little in common. It has also suggested certain new results, particularly in what are called 'decomposition theorems.' But, like the classical theory of finite groups, its primary import seems to be morphological.

I shall start with two examples, the first from the algebra of logic, the second from arithmetic.

In the postulates given earlier [5.2] for the algebra of logic, the basic concepts are a class or set K and two operations $\oplus$, $\odot$. The entire algebra can be (and has been—see Huntington, *loc. cit.* [5.2]) constructed from K and a single binary relation, $\subset$, which may be read 'is included in.' If we think of the elements A, B, C, . . . of K as classes, $A \subset B$ then means 'the class A is included in the class B.' The symbol $\supset$ may be read 'includes'; $A \supset B$ then means 'the class A includes the class B'; and the meanings of $A \subset B$, $B \supset A$ are the same. *Equality*, as in $A = B$, means that everything that is in A is in B, and everything that is in B is in A, or $A \supset B$ and $B \supset A$. These interpretations of the algebra of logic based on $\subset$ instead of on $\oplus$, $\odot$ are of course not necessary. They suffice to show that the algebra is not vacuous, also to suggest interesting things to do in the algebra itself. It is doubtful whether any logician or mathematician ever accomplished much by purely abstract reasoning without some model at the back of his mind. Here the model is class-inclusion.

The entire set of postulates for $\subset$ need not be reproduced. The following will be enough.

(1) For all A in K, $A \subset A$.

(2) If $A \subset B$ and $B \subset A$, then $A = B$.

(3) If $A \subset B$ and $B \subset C$, then $A \subset C$.

(4) $A \oplus B \supset A$, and $A \oplus B \supset B$; and if $C \supset A$ and $C \supset B$, then $C \supset A \oplus B$. Also $A \oplus B = B \oplus A$.

The next are theorems. They connect $\subset$, $\supset$ with $\oplus$.

$$\text{If } A \subset B, \text{ then } A \oplus B = B.$$
$$\text{If } A \supset B, \text{ then } A \oplus B = A$$
$$\text{If } A = B, \text{ then } A \oplus A = A$$

The next are theorems connecting $\subset$, $\supset$ with $\odot$.

$$\text{If } A \subset B, \text{ then } A \odot B = A.$$
$$\text{If } A \supset B, \text{ then } A \odot B = B.$$
$$\text{If } A = B, \text{ then } A \odot A = A.$$

It follows that $A \odot B \subset A$, $A \odot B \subset B$; and if $C \subset A$ and $C \subset B$ then $C \subset A \odot B$. Also $A \odot B = B \odot A$.

These make sense and are intuitively evident in the class-inclusion interpretation of the algebra when $\odot$, $\oplus$ are read as 'intersection,' 'union' respectively. The *intersection* of A,B, denoted here by $[A, B]$ is the *largest* (most inclusive) class whose elements (members) are in both A and B; the *union* (A, B) of A,B is the *smallest* (least inclusive) class whose elements are in A, or in B, or in both. Otherwise expressed, *the intersection $[A, B]$ is the unique most inclusive subclass common to A,B; the union (A, B) is the unique least inclusive superclass containing both A and B.*

It is convenient at this point to draw a distinction between *proper* inclusion, $A \supset B$, $A \neq B$ ('*A* not identical with *B*') and *improper* inclusion, $A \supset B$, $A = B$ ('*A* is identical with *B*'). To indicate that inclusion may be either proper or improper we write $\supseteq$, $A \supseteq B$. 'Equality,' $=$, can be defined in terms of inclusion; $A = B$ if $A \supset B$ and $B \supset A$. (Note that $=$ in this has two meanings.)

The point of what follows in relation to $[A, B]$, (A, B), $\supset$, $\supseteq$ will be seen shortly. We consider a set (or class) S of elements (not necessarily classes) A, B, C, D, D_1, . . . , M, M_1, . . . for which binary relations $>$, $\leq$ are defined and for which the postulates (1), (2) hold. By definition $A \leq B$ and $B \geq A$ are the same.

(1) If A,B,C are any three elements of S such that $A \geq B$ and $B \geq C$, then $A \geq C$.

(2) If A,B are any elements of S, there is an element D of S such that $D \leq A$, $D \leq B$, and such that, if also $D_1 \leq A$, $D_1 \leq B$, then $D_1 \leq D$; there is also an element M of S such that $M \geq A$, $M \geq B$, and such that, if also $M_1 \geq A$, $M_1 \geq B$, then $M_1 \geq M$.

To see that (1), (2) are not vacuous, consider the special case in which the elements A, B, . . . of S are classes as before, $<$ is $\subset$, $\leq$ is $\subseteq$, D is $[A, B]$, and M is (A, B). For these, (1), (2) make sense. If now $(A, (B, C))$ is read, as the definitions demand, as the union of A and (B, C), the latter being the union of B,C, and if $[A, [B, C]]$ is read as the intersection of A and $[B, C]$, the latter being the intersection of B,C, the following theorems are easily provable:

$$[A, B], (A, B) \text{ are uniquely defined.}$$
$$[A, B] = [B, A], (A, B) = (B, A).$$
$$[A, A] = A, (A, A) = A.$$
$$[A, [B, C]] = [[A, B], C].$$
$$(A, (B, C)) = ((A, B), C).$$
$$(A, [A, B]) = A, [A, (A, B)] = A.$$

Among these we recognize our old acquaintances the commutative and associative laws.

For classes the following distributive law is always true.

(3) [*A*, (*B*, *C*)] = ([*A*, *B*], [*A*, *C*]).

There are lattices for which this is not true.

A system defined by (1), (2) is called a *lattice*.

We have interpreted (1), (2) and the theorems in terms of class-inclusion and intersection and union of classes. A radically different interpretation is furnished by the elementary arithmetic of the positive integers 1, 2, 3, ... This will be clear when we recall the definitions of the G.C.D. and L.C.M., not in the traditional form of school arithmetic but in a logically equivalent form which has the advantage that it can be extended to the arithmetic of algebraic numbers [11.5], and indeed much farther. It is an exercise in mere language to see that these revised definitions are equivalent to those of school arithmetic.

If the positive integer *d* divides each of the positive integers *a,b* without remainder, *d* is called a *common divisor* of *a,b*. If every common divisor of *a,b* divides *d*, *d* is called the *greatest common divisor* (G.C.D.) of *a,b*. Thus if *a* = 12, *b* = 18, the common divisors are 1, 2, 3, 6; the G.C.D. is 6, the largest number that divides both 12 and 18, and each of 1, 2, 3, 6 divides 6.

If each of *a,b* divides *m*, *m* is called a *common multiple* of *a,b*. If every common multiple of *a,b* is a multiple of *m*, *m* is called the *least common multiple* (L.C.M.) of *a,b*. The L.C.M. of 12, 18 is 36, the smallest number divisible by both 12 and 18, and every number divisible by both 12 and 18 is a multiple of 36.

The arithmetical interpretation of (1), (2) and their consequences in terms of divisibility, and so on, is: *A*, *B*, *C*, . . . is any class of positive integers; equality, =, is as in common arithmetic; > means 'divides' (without remainder; 2 divides 6, 2 does not divide 5); $<$ means 'is divisible by,' (A, B) is the G.C.D. and $[A, B]$ the L.C.M. of A, B. The detailed verification offers no difficulty. Numerical examples, while proving nothing, make what has been said at least plausible. Thus if $A = 12$, $B = 16$, $[A, B] = 48$, $(A, [A, B]) = (12, 48) = 12 = A$, as it should; $(A, B) = 4$, $[A, (A, B)] = [12, 4] = 12 = A$, again as it should.

❖ 5.6 ❖ *Subrings, Ideals*

Within a field, as we saw [4.2], there may be one or more smaller (less inclusive) fields, called *subfields*, and likewise for a ring, *subrings*. As an example of the latter, the class of all rational numbers a/b, where a, b are integers and $b \neq 0$, evidently is a ring with respect to addition and multiplication as in the arithmetic of common fractions, and this ring contains the subring of all the integers $0, \pm 1, \pm 2, \ldots$, as is clear on restricting b to be 1. The particular subvarieties of a ring called *ideals* have proved of great significance in the general theory of rings, particularly with regard to the structure or morphology of rings.

Ideals entered modern algebra through the theory of algebraic numbers—to be described in Chapter 11—in the 1870s, but it was only in the 1920s and 1930s that their deeper relevance for much of algebra and algebraic geometry was recognized. Ideals are as indispensable as fields, rings, and groups—the last will be described in Chapter 9. Neither of these anticipations is necessary for an understanding of what follows. It will be sufficient here to define ideals for a commutative ring R without, however, assuming the existence of a unity element unless so stated. The elements of R will be denoted by $a, b, c, \ldots, a_1, b_1, c_1, \ldots, x, \ldots$; the 'zero' of R by 0.

An *ideal*, $\mathfrak{s}$, of R is a set of elements of R which is closed under the addition and subtraction of R, and also under multiplication by an arbitrary element of R. In symbols, if a_1, b_1, are in s, so are $a_1 + b_1$, $a_1 - b_1$; for c, in s, and for all x in R, xc, is in s. The set of all elements in s is a subring of R.

Of special importance are the *principal* ideals of R. The principal ideal (a_1) of R is the set of all elements of the form $xa_1 + na_1$, where n is an integer. It is easily seen that (a_1) actually is an ideal. If R has a unity element e, na_1 may be suppressed, as then $na_1 = nea_1$ is the sum of n terms ea_1 each of which is in R. The *zero ideal* is (0); the unit ideal (e) is R itself. A generalization of (a_1) is the ideal $(a_1, \ldots, a_s)$ with a *finite base* $a_1, \ldots, a_s$, consisting of all sums

$$x_1a_1 + \cdots + x_sa_s + n_1a_1 + \cdots + n_sa_s,$$

where $x_1, \ldots, x_s$ are arbitrary elements of R and $n_1, \ldots, n_s$ are integers. As for (a_1), if R has a unity element, $(a_1, \ldots, a_s)$ is the set of all sums $n_1a_1 + \cdots + n_sa_s$.

The next are not intended to be evident. In fact it cost their formulator Dedekind much thought before he hit on them in the 1870s as workable generalizations of multiplication, divisibility, the G.C.D. and the L.C.M. as these notions occur in the arithmetic of the integers 1, 2, 3, ... Dedekind was interested in rings of algebraic integers (described in Chapter 11) in which the rings are commutative and have unity elements. The last is not assumed in what follows. We shall denote ideals of the ring R by capital script letters $\mathfrak{a}$, $\mathfrak{b}$, $\mathfrak{c}$, ...

The product $\mathfrak{a}\mathfrak{b}$ of $\mathfrak{a}$,$\mathfrak{b}$ consists of all sums of products ab, where a is in $\mathfrak{a}$ and b is in $\mathfrak{b}$. At once, from the definition of an ideal, $\mathfrak{a}\mathfrak{b}$ is an ideal, and this multiplication is commutative and associative.

The notion of congruence as defined for the integers goes over formally unchanged to congruences for an ideal modulus $\mathfrak{m}$. The elements a,b of R are defined to be congruent modulo $\mathfrak{M}$ if their difference $a - b$ is in $\mathfrak{M}$, and this is written

$$a \equiv b \bmod \mathfrak{M}.$$

If for the moment $a,b,\mathfrak{M}$ denote integers, this would mean that $a - b$ is divisible by $\mathfrak{M}$. The simple but by no means evident extension of this to rings and ideals held up Dedekind for some time. It will be interesting in passing to follow one clue that suggested the right thing to do.

Consider the fact that 5 divides 20. We are operating for the moment in the ring of ordinary integers. In this ring the principal ideal (5) consists of all integer multiples 0, ± 10, ± 15, ± 20, . . . of 5; the principal ideal (20) consists of all integer multiples 0, ± 20, ± 40, ± 60, . . . of 20. Which of (5), (20) is the 'greater,' in the sense of 'more inclusive'? Clearly (5) includes all of (20), but (20) lacks, among an infinity more, ± 10, ± 30, as these are not multiples of 20.

In symbols, $(5) > (20)$, the principal ideal (5) *includes* the principal ideal (20). But 5 *divides* 20; so it is suggested that as the definition of 'the principal ideal (5) *divides* the principal ideal (20)' we try '(5) *includes* (20),' that is, $(5) > (20)$. This generalizes immediately to commutative rings R.

First, for principal ideals $(a),(b)$ of R. The element $a \neq 0$ of R is defined to divide the element b of R if there is an element c in R such that $b = ac$, and this is written $a|b$, 'a divides b.' It is a simple exercise to prove that if $a|b$, then $(a) > (b)$, (a) includes (b), and conversely, if $(a) > (b)$, then $a|b$. For any ideals $\mathfrak{A}$, $\mathfrak{B}$ of R, $\mathfrak{A} \neq (0)$, '$\mathfrak{A}$ divides $\mathfrak{B}$' is defined to mean $\mathfrak{A} > \mathfrak{B}$: the *more* inclusive 'divides' the *less* inclusive; and the congruence $\mathfrak{A} \equiv (0) \bmod \mathfrak{M}$, or $\mathfrak{M} > \mathfrak{A}$, signifies that all the elements of $\mathfrak{A}$ belong to $\mathfrak{M}$. The *quotient,* $\mathfrak{A}:\mathfrak{B}$ of $\mathfrak{A}$ by $\mathfrak{B}$ is the set of all those elements r of R such that rb is an element of $\mathfrak{A}$ for each element b of $\mathfrak{B}$. It follows that if $\mathfrak{a}:\mathfrak{B} = \mathfrak{c}$, then $\mathfrak{c}\mathfrak{B} \equiv (0)$ mod $\mathfrak{a}$, as for the integers.

The definitions of $(\mathfrak{a}, \mathfrak{B})$, the G.C.D. and $[\mathfrak{a}, \mathfrak{B}]$, the L.C.M. of the ideals $\mathfrak{a}, \mathfrak{B}$ are as follows: $(\mathfrak{a}, \mathfrak{B})$ is the ideal consisting of all sums $a + b$, where a is in $\mathfrak{a}$, b is in $\mathfrak{B}$; $[\mathfrak{a}, \mathfrak{B}]$ is the ideal consisting of all elements common to $\mathfrak{a}$ and $\mathfrak{B}$, that is, their intersection. By comparing these definitions with those given previously [5.5] for the integers, and attending to the definitions here of ideals and congruences of ideals, it is seen that these definitions are *formally* the same as those for the integers. The following analogues of theorems for the integers are easy exercises.

$$\mathfrak{a} (\mathfrak{B}, \mathfrak{c}) = (\mathfrak{a}\mathfrak{B}, \mathfrak{a}\mathfrak{c}).$$
$$\mathfrak{a}\mathfrak{B} \equiv (0) \text{ mod } [\mathfrak{a}, \mathfrak{B}],$$

which states that the product $\mathfrak{a}\mathfrak{B}$ is divisible by the L.C.M. of $\mathfrak{a}, \mathfrak{B}$, or $[\mathfrak{a}, \mathfrak{B}] > \mathfrak{a}\mathfrak{B}$.

$$(\mathfrak{a}, \mathfrak{B})[\mathfrak{a}, \mathfrak{B}] \equiv (0) \text{ mod } \mathfrak{a}\mathfrak{B}.$$

The next among many more connect G.C.D.'s and L.C.M.'s with the definition for quotients given above.

$$\mathfrak{a}:\mathfrak{B} = \mathfrak{a}:(\mathfrak{a}, \mathfrak{B}), \ (\mathfrak{a}:\mathfrak{B}):\mathfrak{c} = (\mathfrak{a}:\mathfrak{c}):\mathfrak{B},$$

and each of the last is equal to

$$\mathfrak{a}:\mathfrak{B}\mathfrak{c}; \ [\mathfrak{a}, \mathfrak{B}]:\mathfrak{c} = [\mathfrak{a}:\mathfrak{c}, \mathfrak{B}:\mathfrak{c}].$$

Step by step, now, the further development follows what was described for congruences and G.C.D., L.C.M. for the integers. There are, of course, differences, as the connotations of the symbols are not the same, but major features of the structures of the two developments are abstractly identical despite their distinct interpretations. Congruence classes are defined as before. If $\mathfrak{a}$ is an ideal of R, all the elements of R fall into *congruence* or *residue classes* mod $\mathfrak{a}$;

the elements r,s of R belong to the same class mod $\mathcal{A}$ if $r \equiv s$ mod $\mathcal{A}$. If the classes are denoted by $\bar{a}$, $\bar{b}$, $\bar{c}$, . . . , the sum $\bar{a} \oplus \bar{b}$, or simply $\bar{a} + \bar{b}$, is defined as the class containing $a + b$ where a is in $\bar{a}$, b is in $\bar{b}$; the product $\bar{a} \odot \bar{b}$, or $\overline{ab}$, is defined as the class containing ab. These congruence classes are a ring with respect to addition, multiplication as just defined. This ring is called the *quotient ring* of R with respect to $\mathcal{A}$, and is symbolized $\mathcal{R}/\mathcal{A}$. This leads to the definition of a *prime ideal* $\mathcal{P}$: if the quotient ring $\mathcal{R}/\mathcal{P}$ is a domain of integrity, the ideal $\mathcal{P}$ is said to be prime. This rather recondite definition is chosen because prime ideals as so defined have some of the distinguishing characteristics (abstractly) of primes in arithmetic.

It may be noted that the ideals form a lattice and indeed one which in general is non-distributive.

❖ 5.7 ❖ *Subfields, Extensions*

I shall take space to mention only two of several notions, *characteristic* and *extension*, that have proved useful in fitting fields into the general scheme of modern abstract algebra. I say 'modern,' because parts of abstract algebra are as old as the pioneering and long-neglected work of the British school of algebraists in the 1830s and 1840s.

It is clear from the definitions that the intersection of all subfields of a field F is a subfield of F, say P, called the *prime* field of F; P is contained in all other subfields of F. Only two types of prime fields are possible. If e is the unity element (with respect to multiplication) of F, e is in P, and hence $e + e$ or $2e$, $e + e + e$ or $3e$, and so on are in P. Either all these elements e, $2e$, $3e$, . . . are distinct, or there is a least positive integer p such that $pe = 0$. In the first case, P must contain all quotients re/se, r,s integers; the set of all these quotients can be mapped onto the set of all rational numbers r/s. The *characteristic* of F is said to be

0 (zero) in this case. In the other case the *characteristic* is p, and it is an easy exercise to prove that p must be prime. Examples of both possibilities occur in systems described earlier, for example in congruence classes of the integers.

In case some high-school student should happen to read the foregoing account of characteristics, I state the easily proved binomial theorem for a commutative field of characteristic p:

$$(a + b)^p = a^p + b^p, \ (a - b)^p = a^p - b^p.$$

Any teacher who has had to induct beginners into the mysteries of elementary algebra will recognize these for $p = 2$, $p = 3$, $p = 5$, . . . as the forbidden 'freshman's delight.'

To explain extension we need the concept of an indeterminate, introduced into algebra by Gauss in 1801. The following is a sufficient account of what an indeterminate, say x, is. The terminology differs from that of Gauss but is equivalent to his. We say that the field H is an extension of F if F is a subfield of H, and further that x is an *indeterminate* over F if the statement

$$ax^n + bx^{n-1} + \cdots + c = 0,$$

where a, b, . . . , c are in F, implies that $a = 0$, $b = 0$, . . . , $c = 0$. For example, an equation in x, $ax^n + bx^{n-1} + \cdots + c = 0$, where the coefficients a, b, . . . , c are real or complex numbers (more generally, elements of a field) is not satisfied by *all* values of x except in the trivial case when each of a, b . . . , c is zero. But if the statement $ax^n + bx^{n-1} + \cdots + c = 0$ is satisfied *only* when $a = 0$, $b = 0$, . . . , $c = 0$, x is an indeterminate. Thus if x is an indeterminate, and

$$ax^2 + bx + c = fx^2 + gx + h,$$

so that

$$(a - f)x^2 + (b - g)x + (c - h) = 0,$$

it follows that $a = f$, $b = g$, $c = h$; that is, corresponding coefficients in the two polynomials are equal. This is one version of the familiar device of 'equating coefficients.'

If a is an element of F not in the prime field P, the set of all rational functions of a with coefficients in P is a field, say $P(a)$, derived from P by the *adjunction* of a. Proceeding from $P(a)$, we may adjoin an element b of F not in $P(a)$, to obtain the field $P(a, b)$, and so on until (with certain reasonable assumptions which I shall omit) we reach F itself by successive adjunctions.

Only two possibilities are open. It will be enough to describe these for the *simple adjunctions*, those obtained by the adjunction of *one* element, say x. The extended field will contain x, x^2, x^3, . . . , and hence all polynomials in x, such as

$$a + bx + cx^2 + \cdots + kx^n,$$

with coefficients a, b, c, . . . , k in F. If no two of these polynomials are equal, then x is an indeterminate over F, and a polynomial can be equal to zero only if all its coefficients are zero. The set of all these polynomials is a domain of integrity. The set of all rational functions (polynomials and their quotients) of x with coefficients in F is a field, say $F(x)$. This field is the smallest containing both F and x. It is said to have been derived from F by *transcendental adjunction* (of x) and is called (a simple) *transcendental extension* of F. The reason for the terminology is that a transcendental number, such as π ($=3.14159$ . . .), is one which satisfies no algebraic equation with complex number coefficients (see also [11.6]). In the second possibility, at least two of the polynomials in x are equal, and hence there is a polynomial $f(x)$ of *lowest degree*, say n, that is equal to zero. By using this polynomial as a modulus (it is irreducible, that is, not a product of polynomials of degrees less than n), all polynomials are reduced in degree to less than n; and two polynomials are equal when and only when they are congruent modulo $f(x)$. The set of all polynomial residues modulo $f(x)$ is readily proved to be a field, which is said to have been derived from F by *algebraic adjunction;* the field is a *simple algebraic extension* of F.

The significance of algebraic extensions is that they provide a means for replacing the fundamental theorem of algebra by an equivalent more in the spirit of algebra than in that of analysis. The fundamental theorem states that every polynomial with complex number coefficients has a root in the field of complex numbers, and it used nearly always to be proved by means of the theory of functions of a complex variable. Older proofs, such as one of several by Gauss, at some point either tacitly or explicitly assumed the continuity of a polynomial, and continuity is foreign to algebra. The strictly algebraic form of the theorem states that, if $f(x)$ is a non-constant polynomial in a field F, there is a simple algebraic extension of F such that $f(x) = 0$ has a root in this extension. Further, any field having this property of the extension must have a subfield equivalent to the extension.

For later reference I must include the distinction between separable and inseparable extensions. We start from the ring $F[x]$ of all polynomials in x with coefficients in the commutative field F. A polynomial in $F[x]$ is *irreducible* if it is not the product of two polynomials, each of at least the first degree in x, with coefficients in F. Otherwise, the polynomial is *reducible*. Let $f(x)$ be an irreducible polynomial of degree at least 1 in x with coefficients in F; that is, $f(x)$ is one of the irreducible polynomials in $F[x]$. If $f(c) = 0$, c is a *root* of $f(x)$. First, if two or more roots of $f(x)$ are equal, $f(x)$ is said to have *multiple roots*. Can this irreducible $f(x)$ have a multiple root? If F has characteristic 0, $f(x)$ has no multiple roots.

If F has characteristic zero and if r is a root of $f(x)$ all of whose roots are distinct (no two equal), r is said to be *separable* with respect to F; in the contrary possibility, r is *inseparable* and likewise for $f(x)$. Finally, a superfield all of whose elements are separable with respect to F is called an *algebraic* superfield, and any other algebraic superfield is called *inseparable*. It may be mentioned that for characteristic p an irreducible (non-constant) polynomial is inseparable if, and only if, it can be written as a polynomial in x^p.

❖ 5.8 ❖ *Skew Fields, Linear Algebras*

If in the postulates for a field [3.1] the postulate for commutativity of multiplication is suppressed, the resulting system, or algebraic variety, is called a *skew field*. We consider the elements of the skew field E with respect to the elements $f_1, f_2, \ldots$ of a field F as coefficients. If in E there are n elements $e_1, e_2, \ldots, e_n$ such that

$$f_1e_1 + f_2e_2 + \cdots + f_ne_n = 0,$$

only when $f_1 = 0$, $f_2 = 0$, $\ldots$, $f_n = 0$, and such that every element of E can be expressed in the form

$$f_1e_1 + f_2e_2 + \cdots + f_ne_n,$$

we call E a *division algebra* with the *finite basis* $e_1, e_2, \ldots,$ e_n. Division algebras are mentioned only in passing, as their theory seems to be pretty well worked out, and in order to describe their properties more fully I should have to introduce too many new definitions. A more general class of algebras is of greater interest to us here, as it was partly responsible for the early stages of the theory of lattices.

We start from a ring R with a unity element. If R contains a field F whose elements as multipliers are commutative with every element of R, and if R has a finite basis e_1, e_2, . . . , e_n, R *is called a finite linear associative algebra of order n with a unity element over F*. The structural properties of this R are implicit in the multiplication table for the units e_1, e_2, . . . , e_n. Since these are a basis, each of the products e_1e_1, e_1e_2, . . . , e_1e_n, e_2e_2, e_2e_3, . . . has an expression of the form $f_1e_1 + \cdots + f_ne_n$, where f_1, . . . , f_n are in F, and it follows by a simple contradiction that the expression is unique. Addition is defined by

$$(f_1e_1 + \cdots + f_ne_n) + (g_1e_1 + \cdots + g_ne_n)$$
$$= (f_1 + g_1)e_1 + \cdots + (f_n + g_n)e_n$$

As an example the following multiplication table defines a linear associative algebra of order 4 with the basis e_1, e_2, e_3, e_4.

$\times$	e_1	e_2	e_3	e_4
e_1	e_1	e_2	e_3	e_4
e_2	e_2	0	0	e_2
e_3	e_3	0	0	e_3
e_4	e_4	$-e_2$	e_3	e_1

It will be noticed that a power of each of e_2, e_3 is zero, $e_2^2 = 0$, $e_3^2 = 0$. There are linear algebras in which some elements x are similarly *nilpotent of order n:* $x^n = 0$, $x^r \neq 0$ for r less than n. Algebraists of the nineteenth century expended much painstaking labor in listing and classifying all linear associative algebras in a small number of units. This taxonomy seems to have been abandoned at $n = 8$, after M. S. Lie (1842–1899), the creator of the theory of continuous groups, more or less abolished the implied problem.

More in line with the ideas I have already described was the fresh start taken in the first decade of the twentieth century. I shall note only a few details which were partly responsible for the theory of lattices [5.5] in the 1930s.

If A is an algebra with a unity element, and B a subalgebra of A with a unity element, $A \cdot B$ here means the set of all products of an element a of A by an element b of B; $B \cdot A$ means the set of all products of an element of B by an element of A. I refer here to the definition of $\subseteq$ as an inclusion relation for sets [5.2]. If now

$$A \cdot B \subseteq B, \text{ and } B \cdot A \subseteq B,$$

B is called an *invariant subalgebra* of A. One theorem, whose proof follows immediately from the definiticns, must suffice here to show the relevance for algebras of inclusions and intersections as previously defined. If B,C are invariant subalgebras with unity elements of the algebra A with a unity element, then $B \cap C = B \cdot C$, and $B \cap C$ is an invariant subalgebra of A whose unity element is the product of the unity elements of B and C.

The next definition relates a linear algebra A to what has been described for ideals [5.6]. It is included because it has proved useful in algebraic geometry. We say that a subalgebra B of A is *invariant* in A if $B \cdot A \subseteq B$, $A \cdot B \subseteq B$, noting that B does not necessarily have a unity element. As the commutative law of multiplication is not assumed here, the previous definition of an ideal must be modified to take account of possible differences of multiplication on the right or on the left by elements of the algebra A for which ideals are defined. Ideals of A are special sets of elements of A closed under addition and subtraction. A *left ideal* $\mathfrak{L}$ is such that $A \cdot \mathfrak{L} \subseteq \mathfrak{L}$; a *right ideal* $\mathfrak{R}$ is such that $\mathfrak{R} \cdot A \subseteq \mathfrak{R}$. If an ideal is both right and left, it is a *two-sided ideal*, or an *invariant subalgebra*. (In fact, a linear algebra is a special kind of ring, and invariant subalgebras are simply ideals. But it was not from this standpoint that the basic theorems of the subject were discovered.) As for ideals discussed previously, congruence for an ideal modulus is defined similarly. If s is a two-sided ideal of A, the elements of A modulo s are an algebra, denoted by A/s—which may be compared with the like for residue classes, or congruence classes, previously discussed [5.3]. The *zero ideal* is o; all its elements are equal to zero. Multiplication of ideals is defined as before, so if s is an ideal of A, s^2, s^3, . . . are defined. An ideal s of A is called *nilpotent* if there is an integer n such that $s^n = o$. The union $(s, \mathcal{I})$, of the ideals $s, \mathcal{I}$ is defined as before, and it can be proved that if $s, \mathcal{I}$ are nilpotent, so is $(s, \mathcal{I})$, also that every nilpotent ideal is contained in a unique maximal (most inclusive) nilpotent ideal, called the *radical* of A. An algebra containing no nilpotent ideal other than o is called *semi-simple;* if it properly contains [5.5] only the ideal o, it is *simple*. It can be proved directly from these definitions that if $\mathfrak{R}$ is the radical of A, $A/\mathfrak{R}$ is semi-simple.

All the preceding notions were current in linear algebra as developed before 1910. They have been recast in the terminology of ideals that became current in the 1920s so that the reader may recognize them should he meet them, as he is likely to do in the rudiments of modern algebra and algebraic geometry. The analogies with other systems described in this chapter are plain. The notions described are among those which gave the *decomposition theorems* of linear algebra relating an algebra to its subalgebras and providing a comprehensive classification of algebras. This offered one of the unmistakable clues to the algebra of structures. In retrospect it seems rather remarkable that this clue was not followed for well over twenty years. But, as has often been remarked of mathematics, ideas seem to choose their own time for being noticed.

For its historical interest I conclude with the multiplication table for the *quaternion units* 1,*i*,*j*,*k*. As already noted [3.2], Hamilton had a long struggle before he imagined a non-commutative multiplication as the key to the algebra of rotations in space of three dimensions.

$\times$	1	i	j	k
1	1	i	j	k
i	i	-1	k	$-j$
j	j	$-k$	-1	i
k	k	j	$-i$	-1

A 'real' quaternion is of the form $a + bi + cj + dk$, where a,b,c,d are real numbers. Real quaternions are the oldest example of a linear associative algebra. This algebra has no divisors of zero; but if the 'coordinates' a,b,c,d may be any complex numbers, then divisors of zero occur. This is most unfortunate for an outstanding problem in the classical theory of numbers. Euler stated (1772) that he believed there are no integers all different from zero such that $a^4 + b^4 + c^4 = d^4$, although he had no idea of a proof. If quaternions with complex coordinates did not have divisors of zero, a proof would be immediate. But they do, and Euler's conjecture remains unsettled.

Hamilton believed that quaternions were the answer to the mathematical physicist's prayer for an algebra of mechanics, optics, and the other physical subjects of his day. But the physicists disagreed. With the advent of the modern (post-1925) quantum theory, however, quaternions were restored to a sort of halfhearted favor when it was noticed that Dirac's equation can be factored into quaternions. The quantum theory has absorbed vast quantities of algebra, digesting some, rejecting others with loathing. For those who care about the aesthetics of mathematical physics apart from its practical necessity of making sense in the world of physical experience, a perfectly gorgeous compost of quaternions and the geometry of that fascinating surface of the fourth degree named after its discoverer E. E. Kummer (1810–1893), the 'wave surface' in a space of four dimensions, may be highly recommended. It seems rather a pity that such beautiful mathematics, designed for physics, should be of no physical significance. But for mathematicians there is the consolation that the only linear associative algebras with real coefficients, having no proper divisors of zero, are the field of complex numbers, its subfields, and the algebra of real quaternions.

Associative algebras do not exhaust either linear algebras or linear associative algebras. Non-associative algebras are as old as 1881, when Cayley came across one in eight units. This algebra is non-commutative. It also, like quaternions, has been used in the quantum theory, but to no great effect. What nuclear physics may demand of algebra before it has split the last nucleus may split algebra into ever smaller and harder fragments and compress algebraists into ever narrower specialists.

Chapter 6

OAKS FROM ACORNS

❖ 6.1 ❖ Transformations

In mathematics it is new ways of looking at old things that seem to be the most prolific sources of far-reaching discoveries. A particular fact may have been known for centuries, and it may have been sterile or of only minor interest all that time, when suddenly some original mind glimpses it from a new angle and perceives the gateway to an empire. What the first flash of intuition sees may take years or even centuries to open up and explore completely, but once a start in the right direction is made, discovery and development go forward at an ever-increasing speed. Such, in outline, appears to have been the evolution of two of the dominating concepts of the mathematics of the nineteenth and twentieth centuries, those of *groups* and *invariants*.

The story begins far back. Distinct traces of the long development are discernible in the work of the Babylonians and the Greeks who, however, never suspected what their regular patterns in tilework and other forms of art meant abstractly, that is, mathematically.[1]

A different approach to the dominating ideas seems to

---

[1] In the December, 1948, *Scientific American*, Professor Dirk J. Struik in his article, Stone Age Mathematics, shows that "the earliest comprehension of number and geometry appears to have been farther back than the time of the Egyptians and Babylonians." The article is illustrated with reproductions of astonishing designs from the decorative art of the neolithic age, the Plains Indians, early Hungary, the primitive tribes of the South Pacific, and the natives of the New Hebrides. The last in particular are remarkable for their intricate symmetries and multiple periodicities. Classical antiquity can show nothing approaching these in harmonious complexity.

have guided the original Moslem algebraists of the ninth to the fifteenth centuries and successive generations of their European followers down to the eighteenth and first two decades of the nineteenth century. But again those who were guided failed to grasp the threads and followed them, if at all, subconsciously.

Regularities and *repetitions* in patterns suggest at once to a modern mathematician the *abstract groups* behind the patterns; and the various *transformations* of one problem, not necessarily mathematical, into another again spell *group* and raise the question *what, if anything, in the problems remains the same,* or invariant, *under all these transformations?* In technical phrase, what are the *invariants* of the *group of transformations?* The current and exact definition of a group must be deferred till Chapter 9, to which the reader may refer now if he wishes. For the moment what has been said must suffice.

When faced with a new problem, mathematicians frequently try to restate it so that it is equivalent to one whose solution is already known. In school algebra, for example, the general equation of the second degree is solved by 'completing the square.' This reduces the general quadratic to one which we can solve at sight. To recall the steps: we solve $y^2 = k$ for y thus, $y = \pm \sqrt{k}$. We then reduce $ax^2 + 2bx + c = 0$, by completing the square, to

$$\left(x + \frac{b}{a}\right)^2 = \frac{b^2 - ac}{a^2},$$

which is of the *same form* as the easy equation $y^2 = k$. In fact, if we now write $y = x + b/a$, $k = (b^2 - ac)/a^2$, we have exactly $y^2 = k$. Notice the expression $b^2 - ac$. It is called the *discriminant* of the equation, because the two roots are real and unequal, equal, or imaginary and unequal according as $b^2 - ac$ is greater than, equal to, or less than zero. A remarkable property of this simple expression, considered in a moment, started the whole vast theory of invariance in both mathematics and science.

Successes such as this were some of the reasons why mathematicians began to study algebraic transformations intensively for their own sake. To illustrate a contributory cause, let us consider two further simple problems, one from elementary algebra, the other from geometry, to see how the comprehensive concept of *invariance* originated.

In $ax^2 + 2bxy + cy^2$, express the x,y in terms of new letters X,Y as follows: $x = pX + qY$, $y = rX + sY$. The result is

$$a(pX + qY)^2 + 2b(pX + qY)(rX + sY) + c(rX + sY)^2$$

Multiply everything out and collect like terms. The result is

$$AX^2 + 2BXY + CY^2,$$

in which A,B,C are the following expressions in terms of a,b,c,p,q,r,s:

$$A = ap^2 + 2bpr + cr^2,$$
$$B = apq + b(ps + qr) + crs,$$
$$C = aq^2 + 2bqs + cs^2.$$

I shall leave it to the reader to verify that the new A,B,C and the old a,b,c are connected by the astonishing relation

$$B^2 - AC = (ps - rq)^2(b^2 - ac).$$

To sum up what has happened, let us write

$$x \rightarrow pX + qY,$$
$$y \rightarrow rX + sY,$$
$$ax^2 + 2bxy + cy^2 \rightarrow AX^2 + 2BXY + CY^2,$$
$$B^2 - AC = (ps - rq)^2 (b^2 - ac).$$

The $\rightarrow$ can be read 'is transformed into.' The indicated transformation of x,y is said to be *linear* (technical term for 'of the first degree') in X and Y. The expression $ps - qr$, which depends only on the coefficients p,q,r,s of the transformation of x,y, is called the *modulus* of this transformation. 'Modulus' here has nothing to do with its use in congruences as in [5.3].

Now look at the summary. It says that $b^2 - ac$ belonging to the original $ax^2 + 2bxy + cy^2$, and $B^2 - AC$, belonging to $AX^2 + 2BXY + CY^2$, differ only by a factor which is the square of the modulus of transformation. For this reason, $b^2 - ac$ is called a *relative invariant* of $ax^2 + 2bxy + cy^2$—'relative,' because $b^2 - ac$ is not *absolutely* unchanged under the transformation. If, however, p,q,r,s are chosen so that $(ps - qr)^2 = 1$, then $b^2 - ac$ and $B^2 - AC$ are *equal and of the same form*, and we say that $b^2 - ac$ is an *absolute invariant* of $ax^2 + 2bxy + cy^2$ under the given linear transformation. This appears to be the first known mathematical instance of such *unchangeableness of algebraic form*.

A mathematician who could look at the relation between $b^2 - ac$ and $B^2 - AC$ and not be at least mildly surprised—provided it was the first time he had seen such a phenomenon—would be little more than an algebraic imbecile. This elementary fact is the acorn, among other things, of one great oak which overshadows modern physics, Einstein's principle of the 'covariance of physical laws,' and it was planted by Lagrange in the late eighteenth century. Cayley, Sylvester, and many others made the acorn grow to the oak in 1846–1897.

Our geometrical example requires no algebra. Consider the shadows cast on a wall by a book as it is turned into various positions. The lengths of the sides of the shadow change as the book is moved. What does *not* change? Try it with a flat mesh of straight wires. The *shadow angles* at which the wires intersect and the *shadow lengths* of the pieces of wire between intersections change in the varying shadows. But an *intersection of two or more wires* remains the same; the shadow wires intersect in the same way as the real wires, and the straight wires *remain straight* in shadow.

The wires represent a simple geometrical configuration of points (intersections) and straight lines. Under the shadow transformation the straightness of the lines is invariant. Further, the intersection of any number of lines is an invariant property, as also is that of the order of any number of intersections lying on one straight line. The shadow is a particular kind of *projection*, like that of a picture on a screen.

The word 'intersection' will be noticed in the preceding paragraph. It has already occurred [5.5], in connection with classes or sets. Is there any significance deeper than a mathematical pun in the two meanings? There is, and it is another of those things that might well have been noticed long before it actually was in 1928. The kind of geometry that deals with properties of configurations invariant under projection is called *projective geometry*. To bring out the point, I quote from the treatise of 1910 on this geometry by O. Veblen (1880–) and J. W. Young (1879–1932). The authors are describing geometry as it appeared to them in 1910. Since then the description of geometry as a whole has been much modified, and no doubt will continue to suffer change for as long as men persist in doing geometry. But what the authors say has retained its force and hints at the deeper significance of 'intersection.'

Geometry deals with the properties of figures in space. Every such figure is made up of various elements (points, lines, curves, planes, surfaces, etc.), and these elements bear certain relations to each other, such as that a point lies on a line, a line passes through a point, two planes intersect, (and so on). The propositions stating these properties are logically interdependent, and it is the object of geometry to discover such propositions and exhibit their logical interdependence.

With lattices behind us, we can see now that the authors were subconsciously thinking of a lattice. Projective geometry (also affine geometry, which I need not discuss) was recast by K. Menger (1902–) in 1928 and independently by G. Birkhoff (1911–), son of the distinguished mathematician G. D. Birkhoff (1884–1944), in 1934 as an instance of lattice algebra. Two associative, commutative operations $+$, $\cdot$, are defined for a system of abstract elements $A, B, \ldots$. These operations admit 'neutral' elements, U the 'universe' (the universal class) and V the 'vacuum' (the null class), such that

$$A + V = A = A \cdot U,$$

for all A in the system, and it is postulated that

$$A + A = A = A \cdot A.$$

All these are familiar to us from the algebra of classes [5.2]. Lines are certain classes of points; points are certain classes of lines—the class of all lines having a common intersection (again!) determines a point. In this algebraization of projective geometry the law of absorption, described in connection with the algebra of classes [5.2], plays the hero, or at least an understudy for him: if $A + B = B$, then $A \cdot B = A$, and conversely for all A,B in the system. The underlying structure, the morphology, of projective geometry was thus revealed as a lattice. Since the initial recasting of projective geometry as a topic in lattice algebra, much detailed work has been done in what may be called lattice geometry.

In contrast to projective geometry, school geometry

(Euclid's) deals almost exclusively with the comparison or measurement of lengths, areas, and angles. For instance, the angle inscribed in a semicircle is a right angle. What becomes of this under projection? It is not invariant, for the circle projects into an ellipse and the right angle loses its 'rightness.'

Properties of geometrical configurations which are *altered* by projection are called *metric*, since they depend upon measurements. Properties *invariant* under projection are called *projective*. This is merely a description of terms and not an exact or full definition. It is sufficient for our purpose, although in passing it may be mentioned that by taking account of points whose coordinates are complex numbers, the whole of metric geometry can be restated more simply as an episode in projection. The common non-Euclidean geometries also come into the shadow picture.

❖ 6.2 ❖ *A Problem in Geometry, Variables Again*

Glancing back at the algebraic example and the geometrical shadows, we see two general problems, one algebraic, the other geometric.

The geometric one is the more easily stated: given any geometrical configuration, to find all those properties of it which are *invariant* (unchanged) under projection.

This is immediately generalized. Why stop with projection, which is only a particular kind of transformation? We might for instance seek all those properties of extensible, flexible surfaces, like sheets of rubber, which are invariant under stretching and bending without tearing. This kind of geometry is called *topology*. It will be described in Chapter 10. The general geometrical problem now is: given any geometric thing—configuration, surface, solid, or whatever can be defined geometrically—and given also a set of transformations of that thing or of the space containing it, to find all those properties of the given thing which are invariant under the transformations of the set.

All this can be translated into the perspicuous symbolic languages of algebra and *analysis*. The last, I recall, may be very roughly described as that department of mathematics which is concerned with *continuous* variables [4.2]. A *variable* is, as its name implies, a mark or letter, say x, which takes on different values successively in the course of a given investigation. For example, the speed of a falling body is not a constant number, say 32 feet per second, but a variable whose numerical value increases continuously from zero (when the body starts to fall) to a greatest speed just as the body strikes the earth. Here the variable is continuous. But variables may be discrete numbers; or they need not be numbers at all, as in symbolic logic [5.2].

I know I should apologize for this very crude description of variables. However, to state fully what a variable is would take a book.[1] And the outcome might be a feeling of discouragement, for our attempts to understand variables would lead us into the morass of doubt [Chapter 20] concerning the meanings of the fundamental concepts of mathematics. I shall ask the reader to trust his feeling for language and let it go at that: a variable is something which changes. A *continuous real variable* passes through all numbers in a given interval, say from zero to 10, or from zero to infinity.

Now, in 1637 R. Descartes (1596–1650) published his epoch-making work on analytical geometry. At one step the whole race of mathematicians strode far ahead of the Greek geometers. To understand the connection between the analytical and algebraical aspects of invariance and the geometrical problem of invariance, it is essential to see

[1] I take comfort in what Weyl says in effect in his profound *Philosophy of mathematics and natural science*, 1949: "Nobody can say what a variable is."

what Descartes did. Accordingly I shall defer to Chapter 10 the further account of geometrical transformations and describe in the following chapter what Descartes did. In brief, he transposed geometry into algebra. That, however, was only the beginning of what followed from his great inspiration. For the present I return to transformations and sketch how matrices entered mathematics.

❖ *6.3* ❖ *Matrices*

I resume the topic of [6.1], with the meaning there defined of linear transformations and the arrow symbol →. One purpose of such transformations, as already noted, is to reduce an unsolved problem to one which either is already solved or is simpler and more approachable than the original. Not to encumber the explanation with extraneous technicalities from geometry, I shall phrase it all in terms of really elementary algebra and eyesight. The letters denote any elements of an abstract field [3.1] but may be thought of as the special instance of any real or complex numbers [4.2]; real numbers are sufficient for visualization. I copy first exactly what Cayley did in 1858, merely modernizing his old-fashioned notation.

Consider first two transformations $A,B,$

$$A: \begin{aligned} x_1 &\to a_{11}y_1 + a_{12}y_2, \\ x_2 &\to a_{21}y_1 + a_{22}y_2; \end{aligned}$$
$$B: \begin{aligned} y_1 &\to b_{11}z_1 + b_{12}z_2, \\ y_2 &\to b_{21}z_1 + b_{22}z_2. \end{aligned}$$

The double suffixes, as in $a_{11}, a_{12}, \ldots$, specify the *row* (first suffix) and *column* (second suffix) in which the *coefficients* $a_{11}, a_{12}, \ldots$ of the transformation A occur, and likewise for B. Thus a_{ij} occurs in a *row i, column j*, and similarly for b_{ij}.

When A,B are applied successively, A *first*, B *second*, written AB, we get

$$AB: \begin{aligned} x_1 &\to a_{11}(b_{11}z_1 + b_{12}z_2) + a_{12}(b_{21}z_1 + b_{22}z_2), \\ x_2 &\to a_{21}(b_{11}z_1 + b_{12}z_2) + a_{22}(b_{21}z_1 + b_{22}z_2), \end{aligned}$$

which, on multiplying out and collecting terms is,

$$\begin{aligned} x_1 &\to (a_{11}b_{11} + a_{12}b_{21})z_1 + (a_{11}b_{12} + a_{12}b_{22})z_2, \\ x_2 &\to (a_{21}b_{11} + a_{22}b_{21})z_1 + (a_{21}b_{12} + a_{22}b_{22})z_2. \end{aligned}$$

This shows how we might pass from x_1,x_2 through y_1,y_2 directly to z_1,z_2. The passage is AB. But, as Cayley must have observed, the formula for AB is not easy to remember. It becomes so when A,B,AB are written schematically in terms of the *arrays*, or *matrices* as Cayley called them, of their coefficients.

$$A: \begin{Vmatrix} a_{11} a_{12} \\ a_{21} a_{22} \end{Vmatrix}, \quad B: \begin{Vmatrix} b_{11} b_{12} \\ b_{21} b_{22} \end{Vmatrix},$$

$$AB: \begin{Vmatrix} a_{11}b_{11} + a_{12}b_{21} & a_{11}b_{12} + a_{12}b_{22} \\ a_{21}b_{11} + a_{22}b_{21} & a_{21}b_{12} + a_{22}b_{22} \end{Vmatrix}.$$

(I use the old double-bar notation, which is less familiar today than that of large parentheses, because it is easier to print, and because I shall presently need parentheses for something else.) Now, if the reader will look closely at these three matrices, A,B,AB, he will see that, given A and B, it is easy to write down AB, the *product* of A and B, *in this order*, without an effort of memory. If the simple rule has eluded anyone, I write down BA, which means B *first*, A *second:*

$$BA: \begin{Vmatrix} b_{11}a_{11} + b_{12}a_{21} & b_{11}a_{12} + b_{12}a_{22} \\ b_{21}a_{11} + b_{22}a_{21} & b_{21}a_{12} + b_{22}a_{22} \end{Vmatrix}.$$

Several things must be noticed. I shall take them one at a time. First, in multiplying as in AB, the rule is '*rows of*

A, the first, onto *columns* of *B,* the second, to give *rows* of the product *AB.*' For example, the first row of *A* may be written $(a_{11},\ a_{12})$; the first column of *B* may be written $\begin{pmatrix} b_{11} \\ b_{21} \end{pmatrix}$; and symbolically, only for a moment,

$$(a_{11},\ a_{12}) \begin{pmatrix} b_{11} \\ b_{21} \end{pmatrix} = a_{11}b_{11} + a_{12}b_{21},$$

which is the element in row 1, column 1 of *AB.* Proceeding in the same way with the *first* row of *A* onto the *second* column of *B* we get

$$(a_{11},\ a_{12}) \begin{pmatrix} b_{12} \\ b_{22} \end{pmatrix} = a_{11}b_{12} + a_{12}b_{22},$$

the element in row 1, column 2 of *AB.* We have now finished with row 1 of *A,* and continue in the same way with the *second* row of *A:*

$$(a_{21},\ a_{22}) \begin{pmatrix} b_{11} \\ b_{21} \end{pmatrix} = a_{21}b_{11} + a_{22}b_{21},$$

the element in row 2, column 1 of *AB;*

$$(a_{21},\ a_{22}) \begin{pmatrix} b_{12} \\ b_{22} \end{pmatrix} = a_{21}b_{12} + a_{22}b_{22},$$

the element in row 2, column 2 of *AB.*

Next I introduce the 'summation convention' which Einstein devised to shorten the formulas of general relativity. This convention was one of the happiest inspirations any algebraist ever had—only Einstein would hardly be flattered at hearing himself called an algebraist; he objects even to being included among the mathematicians.

Notice that in each term of each of the expressions

$$a_{11}b_{11} + a_{12}b_{21},\ a_{11}b_{12} + a_{12}b_{22},$$
$$a_{21}b_{11} + a_{22}b_{21},\ a_{21}b_{12} + a_{22}b_{22}$$

the *second* suffix of a is the *first* suffix of b. For this suffix which the terms have in common we write the letter j and agree that only one term of the sum is to be written, it being understood that the *repeated suffix* (or 'index') j means to take the sum of the indicated terms for $j = 1, 2$. Thus the four sums displayed above are written

$$(a_{1j}b_{j1}), \ (a_{1j}b_{j2}),$$
$$(a_{2j}b_{j1}), \ (a_{2j}b_{j2}),$$

where the parentheses direct us to sum the typical term enclosed in them over the values $j = 1, 2$.

The next thing in order is the concept of a *vector*, which I shall describe in its general form (*n components* $x_1, \ldots, x_n$ instead of only two as in the explanation so far). A vector of *order n* is here a set of n elements (of a commutative field F, say) $x_1, x_2, \ldots, x_n$ arranged in a definite sequence x_1 first, x_2 second, $\ldots, x_n$ nth, and is written ($x_1, x_2, \ldots, x_n$). Elements of F are called *scalars*. For brevity the vector just written is denoted by (x), and by definition the product $c(x)$ of x by any scalar c is the vector ($cx_1, cx_2, \ldots, cx_n$). The vector (y) is ($y_1, y_2, \ldots, y_n$). Equality of vectors, $(x) = (y)$, is defined to mean simultaneous equality of similarly numbered components, $x_1 = y_1, x_2 = y_2, \ldots, x_n = y_n$. It is clear that this equality satisfies all the postulates of an equivalence relation [5.3]. The scalar product $(x)c$ is by definition the vector ($x_1c, x_2c, \ldots, x_nc$), and since the field F was assumed to be commutative, this is ($cx_1, cx_2, \ldots, cx_n$), so that $c(x) = (x)c$. *Addition* of vectors $(x) + (y)$ is defined by

$$(x) + (y) = (x_1 + y_1, x_2 + y_2, \ldots, x_n + y_n).$$

The *zero* vector of order n is $(0, 0, \ldots, 0)$, each of whose n components is zero. With these definitions it is easily seen that the set of all vectors of order n is closed under addition and its inverse, subtraction, defined by $(x) - (y) = (x_1 - y_1, x_2 - y_2, \ldots, x_n - y_n)$, and under scalar multiplication, as in $c(x)$. Notice particularly that a general *multiplication* is *not* defined for *these* vectors. But a very *special kind* of multiplication *is* defined. The *scalar product* of the vectors (x), (y), is the *scalar* (element of F)

$$x_1 y_1 + x_2 y_2 + \cdots + x_n y_n,$$

and we shall (unorthodoxly) write this scalar product as $(x) \cdot (y)$ for simplicity. The algebraic rules just stated for vectors (x), (y), $\ldots$ define a *linear vector space*. Here we are interested only in the immediate application to matrices of what has been said. With the illustration of multiplication for matrices of *two* rows and *two* columns, or '2×2' ('two by two') matrices, as a guide I shall pass at once to the immediate generalization.

An '$m \times n$ matrix,' read 'm by n matrix,' denoted by $|a_{ij}|$, in a field F, is an array of mn elements of F arranged in the m by n array next indicated (m rows, n columns),

$$\begin{vmatrix} a_{11} & a_{12} & \cdots & a_{1n} \\ a_{21} & a_{22} & \cdots & a_{2n} \\ \cdot \cdot \cdot & & \cdot \cdot \cdot & \cdot \cdot \cdot \\ a_{m1} & a_{m2} & \cdots & a_{mn} \end{vmatrix},$$

so that a_{ij} is the element in *row i, column j*. So far nothing much has been said. The following definitions (or postulates) supply what is needed to make matrices fruitful. In what immediately follows, $|a_{ij}|$, $|b_{ij}|$ are $m \times n$ matrices. (Notice the $m \times n$, which cannot be replaced by $n \times m$ unless $m = n$, because the *first* number, m, refers to *rows*, the *second*, n, to *columns*.)

The two $m \times n$ matrices

$$\|a_{ij}\|, \|b_{ij}\|, \, i = 1, 2, \ldots, m, \, j = 1, 2, \ldots, n,$$

are defined to be *equal*, $\|a_{ij}\| = \|b_{ij}\|$, if, and only if, $a_{ij} = b_{ij}$ for all i, j as above.

The *sum* of these two matrices is by definition the matrix $\|a_{ij} + b_{ij}\|$; that is, the element in row i, column j of the sum of the matrices is the sum of elements in row i, column j of the added matrices.

The *scalar product* $c\|a_{ij}\|$ of the scalar (element of F) c and the matrix $\|a_{ij}\|$ is by definition the matrix $\|ca_{ij}\|$.

The *zero* $m \times n$ matrix is by definition the matrix of m rows and n columns each of whose elements is the zero element of F.

The next thing is to look at an $m \times n$ matrix in each of two different ways, first as an $m \times 1$ matrix—that is, as a matrix of m *rows* and 1 *column*, each of whose elements is a vector of order n—and second as a $1 \times n$ matrix—that is, as a matrix of 1 row and n columns, each of which is a vector, written vertically instead of horizontally, of order m. Instead of trying to make sense of all these words the reader may see on glancing at the figure what is meant:

$$\begin{Vmatrix} a_{11} & a_{12} & \cdots & a_{1n} \\ a_{21} & a_{22} & \cdots & a_{2n} \\ \cdots & \cdots & & \cdots \\ a_{m1} & a_{m2} & \cdots & a_{mn} \end{Vmatrix} = \begin{Vmatrix} R_1 \\ R_2 \\ \cdot \\ \cdot \\ \cdot \\ R_m \end{Vmatrix} = \|C_1 \quad C_2 \quad \cdots \quad C_n\|,$$

where R_i $(i = 1, \ldots, m)$ stands for *row* i, written horizontally as the vector $(a_{i1}, a_{i2}, \ldots, a_{in})$, and $C_j(j = 1,$

..., n) stands for *column j*, written vertically as the vector

$$\begin{pmatrix} a_{1j} \\ a_{2j} \\ \cdot \\ \cdot \\ \cdot \\ a_{nj} \end{pmatrix}.$$

Next, suppose we have the above $m \times n$ matrix, and wish to write an $n \times m$ matrix—notice the interchange of m,n. As m,n were unrestricted in the above definition, we know how the $n \times m$ matrix is to be written. Say its elements are b's:

$$\begin{vmatrix} b_{11} & b_{12} & \cdots & b_{1m} \\ b_{21} & b_{22} & \cdots & b_{2m} \\ \cdot & \cdot & \cdots & \cdot \\ b_{n1} & b_{n2} & \cdots & b_{nm} \end{vmatrix} = \begin{Vmatrix} R'_1 \\ R'_2 \\ \cdot \\ \cdot \\ R'_n \end{Vmatrix} = \begin{Vmatrix} C'_1 & C'_2 & \cdots & C'_m \end{Vmatrix},$$

where the accents, as in R'_1, C'_1, and so on, are supplied to distinguish this matrix from the other.

Now return to our initial example of a 2×2 matrix and recall the rule of multiplication, 'rows of the first onto columns of the second.' Call the a,b matrices displayed above A,B. The *product AB, in this order*, of A,B is defined to be the matrix whose element in *row i column j* is the *inner product* $R_i \cdot C'_j$ of the vectors R_i, C'_j; that is,

$$AB = \lfloor R_i \cdot C'_j \rfloor,$$
$$R_i \cdot C'_j = a_{ir} \cdot b_{rj},$$

where the repeated index r as already explained means the

summation over $r = 1, 2, \ldots, n$. Actually we have applied the 'rows onto columns' rule to AB written in the form

$$\begin{Vmatrix} R_1 \\ R_2 \\ \cdot \\ \cdot \\ \cdot \\ R_m \end{Vmatrix} \begin{Vmatrix} C'_1 & C'_2 & \cdots & C'_m \end{Vmatrix},$$

that is, as an $m \times 1$ matrix multiplied by a $1 \times m$ matrix, to get the $m \times m$ matrix

$$\begin{Vmatrix} R_1 \cdot C'_1 & R_1 \cdot C'_2 & \cdots & R_1 \cdot C'_m \\ R_2 \cdot C'_1 & R_2 \cdot C'_2 & \cdots & R_2 \cdot C'_m \\ \cdots & \cdots & \cdots & \cdots \\ R_m \cdot C'_1 & R_m \cdot C'_2 & \cdots & R_m \cdot C'_m \end{Vmatrix}.$$

An $m \times m$ matrix is called a *square* matrix of *order* m. Square matrices in several respects are more interesting than the general rectangular $m \times n(n \neq m)$ matrices. In all that follows we shall attend only to square matrices. The *unit* matrix I_m of order m has 1's down its main diagonal (top left corner to bottom right corner) and 0's everywhere else. It is immediate that for any matrix A of order m,

$$I_m A = A I_m = A.$$

The zero matrix O_m of order m has all its elements zero. The suffix m may be dropped when it is understood that all the matrices in a specific context have the same order m.

Some readers may be interested in verifying that the set of all square matrices $A, B, C, \ldots, I, O, \ldots$ of order m is a ring [5.3] with unity element I. The negative, $-A$, of A is the matrix whose elements are the negatives of those of

A, so that $A + (-A)$, or simply $A - A$, is O. It is to be shown that the associative laws of addition and multiplication hold,

$$A + (B + C) = (A + B) + C, A(BC) = (AB)C,$$

also the distributive laws,

$$A(B + C) = AB + AC, (B + C)A = BA + CA,$$

where it is necessary to state both because multiplication of matrices is not in general commutative. The easiest way is to use the summation convention for inner products.

I shall now describe what seems to me an extraordinary theorem discovered by Cayley, and I shall follow him in exhibiting it first for matrices of order 2: *a matrix satisfies its characteristic equation.* For the matrix

$$A = \begin{vmatrix} a & b \\ c & d \end{vmatrix}$$

this means that, *identically*,

$$A^2 - (a + d)A + (ad - bc)I = 0,$$

where (since m here is 2),

$$I = \begin{vmatrix} 1 & 0 \\ 0 & 1 \end{vmatrix}, O = \begin{vmatrix} 0 & 0 \\ 0 & 0 \end{vmatrix}.$$

To verify this we do the indicated calculations:

$$A^2 = \begin{vmatrix} a & b \\ c & d \end{vmatrix}\begin{vmatrix} a & b \\ c & d \end{vmatrix} = \begin{vmatrix} a^2 + bc & ab + bd \\ ca + dc & cb + d^2 \end{vmatrix};$$

$$-(a + d)A = -(a + d)\begin{vmatrix} a & b \\ c & d \end{vmatrix} = \begin{vmatrix} -a^2 - ad & -ab - db \\ -ac - dc & -ad - d^2 \end{vmatrix};$$

$$(ad - bc)I = (ad - bc)\begin{vmatrix} 1 & 0 \\ 0 & 1 \end{vmatrix} = \begin{vmatrix} ad - bc & 0 \\ 0 & ad - bc \end{vmatrix}$$

Adding these three matrices we get, for

$$A^2 - (a + d)A + (ad - bc)I,$$

the matrix

$$\begin{vmatrix} a^2 + bc - a^2 - ad + ad - bc & ab + bd - ab - db + 0 \\ ca + dc - ac - dc + 0 & cb + d^2 - ad - d^2 + ad - bc \end{vmatrix};$$

which is

$$\begin{vmatrix} 0 & 0 \\ 0 & 0 \end{vmatrix},$$

or O as stated.

Cayley does not say what suggested this to him. The corresponding equation for matrices of order 3 is too long to write out in full here. Cayley wrote it out and verified it, then boldly stated that the similar equation for matrices of order m holds. He offered no proof, and it is not easy to give one unless you have been shown how.

To state the general theorem I shall have to assume that the reader is acquainted with the expansion of a determinant as given in the second course of school algebra. The *determinant* of the matrix A,

$$A = \begin{vmatrix} a_{11} & a_{12} & \cdots & a_{1m} \\ a_{21} & a_{22} & \cdots & a_{2m} \\ \cdots & \cdots & \cdots & \cdots \\ a_{m1} & a_{m2} & \cdots & a_{mm} \end{vmatrix},$$

is

$$\begin{vmatrix} a_{11} & a_{12} & \cdots & a_{1m} \\ a_{21} & a_{22} & \cdots & a_{1m} \\ \cdots & \cdots & \cdots & \cdots \\ a_{m1} & a_{m2} & \cdots & a_{mm} \end{vmatrix}.$$

Now subtract x from each element in the principal diagonal of this determinant,

$$\begin{vmatrix} a_{11} - x & a_{12} & \cdots & a_{1m} \\ a_{21} & a_{22} - x & \cdots & a_{2m} \\ \cdots & & \cdots & \cdots \\ a_{m1} & a_{m2} & \cdots & a_{mm} - x \end{vmatrix},$$

expand this determinant and arrange the result as a polynomial in x,

$$x^m + c_1 x^{m-1} + c_2 x^{m-2} + \cdots + c_m.$$

Then it can be proved that

$$A^m + c_1 A^{m-1} + c_2 A^{m-2} + \cdots + c_m I_m = O_m.$$

For those who have studied determinants I shall sketch one of the simplest and most useful applications of matrices, that to the solution of a system of linear equations. A matrix whose determinant is not zero is called *non-singular*. If the matrix A of order m is non-singular, it can be shown without difficulty that there is a unique matrix, denoted by A^{-1} and called the *inverse* of A, such that

$$A A^{-1} = A^{-1} A = I_m.$$

The system of equations in $x_1, \ldots, x_m$,

$$a_{11} x_1 + a_{12} x_2 + \cdots + a_{1m} x_m = c_1,$$
$$a_{21} x_1 + a_{22} x_2 + \cdots + a_{2m} x_m = c_2,$$
$$\cdots \qquad \qquad \cdots \qquad \qquad \cdots$$
$$a_{m1} x_1 + a_{m2} x_2 + \cdots + a_{mm} x_m = c_m,$$

can be written in the much condensed form

$$a_{ij} \cdot x_j = c_j,$$

where it is understood that each of i, j ranges over 1, 2, . . . , m, and $a_{ij} \cdot x_j$ is the scalar (or inner) product already defined (j is the repeated index indicating a summation). The *matrix of the system* is A as written out above. The condensed form can be written

$$A \begin{vmatrix} x_1 \\ x_2 \\ \cdot \\ \cdot \\ \cdot \\ x_m \end{vmatrix} = \begin{vmatrix} c_1 \\ c_2 \\ \cdot \\ \cdot \\ \cdot \\ c_3 \end{vmatrix}.$$

If A is non-singular, it has an inverse, A^{-1}. Suppose A is non-singular. Multiply the preceding equation throughout, *on the left* by A^{-1}. Then

$$\begin{vmatrix} x_1 \\ x_2 \\ \cdot \\ \cdot \\ \cdot \\ x_m \end{vmatrix} = A^{-1} \begin{vmatrix} c_1 \\ c_2 \\ \cdot \\ \cdot \\ \cdot \\ c_m \end{vmatrix}.$$

All this can be further condensed, but perhaps the above is a sufficient hint of the final compression.

One reason for looking at a system of linear equations in this way is that some of the calculating machines deliver an inner product of two vectors by a single operation. I believe it was A. S. Eddington (1882–1944) who first gave this condensed way of manipulating a system of linear equations. It was suggested by the tensor algebra of general relativity. Eddington worked it out to aid his friends in mathematical statistics, who had asked for something less bulky than the standard presentation of the textbooks.

Almost any example written down at random will show that multiplication of matrices is only exceptionally commutative. For example,

$$\begin{vmatrix} 1 & 2 \\ 3 & 4 \end{vmatrix} \begin{vmatrix} 5 & 7 \\ 6 & 8 \end{vmatrix} = \begin{vmatrix} 17 & 23 \\ 39 & 53 \end{vmatrix},$$

$$\begin{vmatrix} 5 & 7 \\ 6 & 8 \end{vmatrix} \begin{vmatrix} 1 & 2 \\ 3 & 4 \end{vmatrix} = \begin{vmatrix} 26 & 30 \\ 38 & 44 \end{vmatrix}.$$

Another peculiarity of matrices is that they may have proper divisors of zero [5.3], where 'zero' here means the zero matrix of the relevant order. Thus if a, b, c, d are any elements not all zero of the field F such that $ad = bc$,

$$\begin{vmatrix} a & b \\ c & d \end{vmatrix} \begin{vmatrix} -b & -d \\ a & c \end{vmatrix} = \begin{vmatrix} 0 & 0 \\ 0 & 0 \end{vmatrix},$$

and neither factor is the zero matrix.

For its historical interest I shall conclude these examples with the matrix representation of Hamilton's quaternion units $1,i,j,k$ [5.8], leaving it to the reader to verify that the matrices shown actually do satisfy the defining equations [5.8] for these units when the scalar 1 is replaced by the 2×2 unit matrix I:

$$I = \begin{vmatrix} 1 & 0 \\ 0 & 1 \end{vmatrix}, \ i = \begin{vmatrix} \sqrt{-1} & 0 \\ 0 & -\sqrt{-1} \end{vmatrix}, \ j = \begin{vmatrix} 0 & 1 \\ -1 & 0 \end{vmatrix},$$

$$k = \begin{vmatrix} 0 & \sqrt{-1} \\ \sqrt{-1} & 0 \end{vmatrix}.$$

For example,

$$i^2 = j^2 = k^2 = ijk = -I,$$
$$ik = -j, \ kj = -i, \ ki = j, \ ji = -k, \ jk = i.$$

Looking back, we may see that this sample of the algebra of matrices has illustrated what it is possible to get out of a well-devised mathematical notation. As P. S. Laplace (1749–1827) observed, half the battle in mathematics is the invention of a good notation. Gauss on the other hand rather sourly remarked (in Latin) that mathematics is more concerned with notions than with notations. Possibly neither was wrong.

'Notational' or 'formalistic' mathematics may not be very deep, but it certainly is broad, occasionally broader even than excessively thin-spread notional mathematics. The algebra of matrices, for example, is now of vast extent, and its numerous applications range from mathematical physics to the theory of statistics, in both of which, if not indispensable, it is at least a useful aid. In pure mathematics this algebra is as necessary as is that of a field. And it all came out of Cayley's apparently trivial remark that a linear transformation can be represented by the array of its coefficients. This history recalls a story about V. Hugo (1802–1885). He started writing one of his novels with a full bottle of ink at his elbow. When the last word was written the last drop of ink was gone. He wanted to call his novel "What there is in a bottle of ink," but his publishers objected. Hugo was too modest; he overlooked his brains.

❖ 6.4 ❖ A Suggestion to the United Nations

The concept of invariance became of first-rate mathematical importance, particularly in algebra and algebraic geometry, about the middle of the nineteenth century. It became of first-rate scientific importance in 1915. In that year, we recall, poison gas was first used as a weapon of Christian warfare and Einstein had practically completed his general theory of relativity and gravitation. Thirty years later, in 1945, Einstein's equation, $E = mc^2$, in which E is (in the appropriate units) the amount of energy in a mass m and c is the enormous number representing the speed of light, that came out of the special theory of relativity of 1905, had matured in the atom bomb that blasted the heathen city of Hiroshima off the infidel map. Invariance had a very considerable part in the mathematics and physics that produced $E = mc^2$, as any interested reader may check for himself. From whatever vantage point we view it, human progress certainly is remarkable. (It may be true that $E = mc^2$ was not directly responsible for atomic fission, but that it was one of about twelve historically significant steps toward Hiroshima is claimed by competent authorities.) Possibly even more spectacular progress than that provoked by the special theory of relativity will eventuate from applications to national defense—and international suicide—of the general theory of relativity, one of the strikingly beautiful scientific applications of invariance. We shall look at it later [10.3]. Likewise for the quantum theory. As early as 1946 cosmic rays were being sniffed at for their military ('defense') possibilities.

So far as invariance is concerned in all this I may refer to the historical origin of the entire theory as already described [6.1]. The theory originated, as we saw, in Lagrange's innocuous observation that the discriminant of a binary quadratic 'form' is invariant under linear homogeneous transformations of the variables, or indeterminates [5.7], in the form. Lagrange discovered this while investigating perfectly useless problems in the theory of numbers. His discovery might well have been made, and indeed may have been made, long before he published it, by some inquisitive child playing with quadratic equations in the first six months of school algebra. Lagrange himself did not see the tremendous importance of his discovery. Nor did his successors until the 1840s. Nevertheless, the germ of a revolutionary idea is now plainly apparent in what he did.

Is not the lesson of all this evident? Should not the

United Nations prohibit the teaching of quadratic equations
in the schools of the world? Possibly it is too late to ban
quadratics, even for the Eskimos. But beyond quadratics
for more advanced peoples there are cubics, quartics, and
so on, and from there on all the mathematics, pure and
applied, from the calculus to quantum mechanics and relativity and the devil only knows what else. The remedy here
is so evident that it need only be mentioned: ban all the
books on mathematics and science. And that, whether you
believe it or not, is exactly what has been seriously proposed by overheated humanists and panicky pulpiteers.

✤ 6.5 ✤ *Invariance in Nature*

'Invariance' means unchangedness, and its mathematical
theory is concerned with discovering whatever, if anything,
other than human perversity, it may be that remains unchanged in mathematical expressions when the variables in
those expressions are replaced by functions of themselves
or of other variables.

The general concept of invariance is older than its mathematical formulation. In theology, for example, we encounter
that appalling phrase, "The same yesterday, today, and
forever." This probably is the earliest authentic example
of an absolute invariant for all transformations of the familiar space-time of general relativity. I pass on to secular
manifestations of invariance.

The total amount of matter in the universe was an invariant for the chemists of the nineteenth century. The
physicists of the same period assumed that the total amount
of energy in the universe was an invariant. Today the total
amount of mass plus energy is an invariant. The first of
these assumptions was called the 'law' of the conservation
of matter, the second the 'law' of the conservation of
energy. Both were extremely fruitful hypotheses, and both

were inferences from innumerable observations. Both raised hell with civilization as exemplified in coal mines, steam engines, industrial revolutions, child labor, and so on. But let this pass—if indeed it has not already passed.

The classic experiment that begot the first law—or was at least its grandfather—was A. L. Lavoisier's (1743–1794) with the oxidation of mercury, described in the schoolbooks on chemistry.

Lavoisier is famous for more than his chemistry. He inspired one of the stupidest pronouncements in history. When in the French Revolution Lavoisier's friends pleaded for his life on the ground that he was a great scientist, the proletarian judge declared, "The People have no need of science." Lavoisier lost his head. Lagrange's comment on this episode may be of interest: "It took them only a moment to cause this head to fall, and a hundred years perhaps will not suffice to produce its like."

A less costly demonstration than Lavoisier's salvages all the solids and invisible gases from a burning candle, weighs them, and shows that although the candle has vanished none of its substance has been lost. Quite the contrary appears at first to have been the case. The fully liberated soul of the candle seems to weigh more than the candle itself. This is easily shown to be an illusion by recovering the oxygen contributed by the air to the products of combustion. On restoring to the air the things that were its, and to the candle the things that were its, the arithmetic comes out right: the total amount of matter is the same after burning as it was before.

In the case of energy it was far more difficult to prove invariance experimentally. Energy of motion, electrical energy, and thermal energy were harder than mass to measure accurately. But again the sum came out constant, even in such delicate measurements as those necessary to show that part of the energy of a sound wave is dissipated in heat, causing a rise in the temperature of the agitated air. It has even been calculated, I believe, that the heat generated by all the oratory of a presidential campaign is insufficient to fry a gnat's egg. Nevertheless it is from quantities of this all but inconceivably small order that we recover the profoundest secrets of nature, human and other.

❖ 6.6 ❖ *Sylvester's Prevision*

The theories of matrices and (algebraic) invariants are only a small sample of the algebra that recent science has found it profitable to apply. Modern algebra and modern mathematics in general contain a great deal more that has not yet passed into scientific circulation. To judge by the past, much of this treasure will also some day become part of the medium of exchange between scientists and nature.

Limitations of space forbid more than a passing mention of the remarkable prophecy (1878)—'hunch' is perhaps juster—of that eloquent algebraist Sylvester, who once declared that an eloquent mathematician is as rare as a talking fish. We shall let him tell in his own words how his inspiration came to him.

Casting about as I lay awake in bed one night, to discover some means of conveying an intelligible conception of the objects of modern algebra to a mixed society mainly composed of physicists, chemists, and biologists, interspersed with only a few mathematicians, . . . I was agreeably surprised to find, all of a sudden, distinctly pictured on my mental retina a chemico-algebraical image, serving to embody and illustrate the relations of those derived algebraical forms to their primitives and to each other which would perfectly accomplish the object I had in view.

If the reader will glance back at the formula in [6.1] expressing the invariance of $b^2 - ac$, he will see the simplest of 'those derived algebraical forms' of which Sylvester speaks.

This was derived from the homogeneous[1] form of the second degree in two variables x,y by a homogeneous transformation of the first degree on x,y. If the same transformation be applied to the homogeneous form of degree n, in two variables, invariant expressions (involving only the coefficients a_0, a_1, . . . , a_n, or both the variables x,y and the coefficients) are discovered, and these are connected by certain algebraic relations, 'syzygies,' as Sylvester, thinking possibly of the moon, called them. The structure of these relations mimics in a curious sort of parody the structural formulas of chemical compounds. From this Sylvester predicted that the theory of algebraic invariants would provide a clue through the intricacies of chemical valence.

The prediction was ignored, except by a few mathematicians, including Clifford, who further developed the algebra but who were incapable of connecting their formulas in any significant manner with chemistry. In the early 1900s an extremely scholarly but somewhat unimaginative mathematician took great pains in his article on the theory of invariants for the colossal German encyclopedia of mathematics to ridicule Sylvester's prophecy. He was quite emphatic about the fantastic nonsense of the 'chemico-algebraical' theory of valence which Sylvester had imagined.

In 1930 it was found that Sylvester had prophesized correctly (see also [9.5]).

---

[1] $a_0x^n + a_1x^{n-1}y + a_2x^{n-2}y^2 + \cdots + a_{n-1}xy^{n-1} + a_ny^n$, in which each term is of degree n in x and y, is called homogeneous of degree n in x and y.

Chapter 7

PICTORIAL THINKING

❖ *7.1* ❖ *Graphs*

The physicist and engineer Lord Kelvin (William Thomson, 1824–1907), who in the late 1850s was scientific consultant for the first transatlantic telegraph cable, once remarked that a single curve, "drawn in the manner of the curve of prices of cotton," can depict all that the ear can possibly hear in the most complicated musical performance. This is so, as may easily be imagined by anyone who has inspected a record for a mechanical piano, or better who has visualized how the spiral grooves on the phonograph recording of a symphony would look if drawn out into a line. And this possibility, Kelvin continued, was to him a wonderful proof of the potency of mathematics. We shall note several even more wonderful proofs of the same potency as we proceed. Here the curve depicting a symphony is a sufficient suggestion.

From market reports and charts of the ever-deepening trough of the depression of the 1930s, graphical representations of masses of involved statistics, comprehensible at a glance when plotted as curves, became distressingly familiar to every reader of the newspapers. But as late as the early 1900s such intuitively clear graphs were not a commonplace to even the well-educated, and H. G. Wells (1866–1947) in one of his numerous projects for civilizing society urged that the art of reading graphs be made compulsory in all elementary education. Presently the more progressive pedagogues caught Wells's enthusiasm for graphs. Soon it was impossible for any unsuspecting child to open a book on arithmetic without being confronted with jiggly rainfall histories and sinuous tide tables. In algebra the infection was even more acute, until everything from simultaneous equations to complex numbers was 'graphed' silly. Wells's project had materialized. But, disconcertingly enough, the children hated graphs. The common-sense reason for this should have been plain to anyone with only a slight knowledge of mathematical history. It took over two thousand years for great mathematicians to hit upon the simple but sophisticated idea underlying the graphical representation of numerical data. So why should it seem natural or desirable to a child? Yet today, thanks partly to Wells, educated adults in our print-bedeviled society understand graphs whether they want to or not. It is not clear, however, that society is less uncivilized than it was before the great plague of graphs.

That graphs are extremely useful in everyday life no one who can still see would deny. Of themselves graphs are also indispensable in science, as may be checked by glancing through any book on physics or chemistry or biology. But their greater scientific significance is in what they inspired. From the graphical representation of numerical data evolved a usable conception of 'space.' Descartes' *analytic geometry* of 1637 prepared the way for the 'space' of any number of dimensions which is now a commonplace in applied mathematics from mechanics and the physics of gases to relativity and intelligence testing.[1]

It was Descartes who took the first and decisive step toward transposing what we ordinarily think of as *geometry* into *relations between numbers*. After this initial step the

[1] Mathematical readers will recall the 'realization' of certain of the advanced parts of the theory of correlation as spherical trigonometry in a space of n dimensions.

next, in retrospect, should have been easy and might well have followed within a generation, or at latest by 1700. Actually it was delayed for over two centuries. The geometry of a space of more than three dimensions was effectively created only in 1844. Almost concurrently there developed the extremely fruitful custom of translating algebra and analysis—the mathematics of continuous change—into the graphic language of 'space' by using the traditional vocabulary of geometry.

As these two complementary themes will recur repeatedly in the sequel, we must see clearly the very simple thing Descartes did. In describing it I shall not follow him exactly but shall incorporate certain immediate amplifications supplied by his successors.

❖ 7.2 ❖ What Descartes Invented

To anyone who has lived in a modern American city (except Boston) at least one of the underlying ideas of Descartes' analytic geometry will seem ridiculously evident. Yet, as remarked, it took mathematicians all of two thousand years to arrive at this simple thing.

Imagine a city laid out in avenues running east to west and streets running north to south. Any address, say 5124 South 81st Street, is instantly visualized. The mathematics of this convenient scheme of locating points in a plane, *by assigning a pair of numbers to each point*, provides the basis for the *algebraic* description of any straight line or of any plane curve no matter how complicated. The meat of it all is in Figure 7.

To locate any point in a plane, lay down two straight lines XOX', YOY', intersecting at right angles at O. (The reference lines XOX', YOY' could intersect at any angle, but there is no essential gain in generality in choosing other than a right angle.) These lines may be laid down anywhere in the plane. The 'positions' of all points in the plane are to be referred to this *reference system*, or *coordinate system*. The point O is called the *origin*, and the lines XOX', YOY' the *axes*, of coordinates. Distances measured to the *right* of YOY' in the direction OX are (arbitrarily) called positive, and are *represented* by *positive numbers;* distances to the *left* of YOY' in the direction OX' are negative, and are *represented* by *negative* numbers. Similarly distances measured

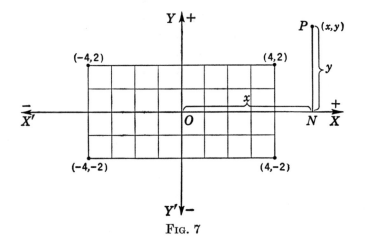

Fig. 7

in the direction OY *above* XOX' are *positive;* distances measured in the direction OY' *below* XOX' are *negative.*

A distance measured parallel to XOX', in either direction, is denoted by x, and is a *positive* or a *negative number* according as the distance is measured to the *right* or the *left* of YOY'.

A distance measured parallel to YOY', in either direction, is denoted by y, and is a *positive* or a *negative number* according as the distance is measured *above* or *below* XOX'.

For these reasons it is convenient to call XOX' the x axis, and YOY' the y axis.

The axes divide the plane into four *quadrants*: the *first* is *XOY*, the *second* is *YOX'*, the *third* is *X'OY'*, and the *fourth* is *Y'OX*. That is, in numbering the quadrants we go *counterclockwise*, which is by definition the *positive* direction of rotation.

Referring to the figure, we *label* any point *P* in the plane determined by the axes by assigning to the point a pair, (x, y), of numbers *x,y*, the first of which, *x*, is the distance (positive or negative) that *P* lies beyond *YOY'* (as already described), and the second of which, *y*, is the distance that *P* lies beyond *XOX'*. In the figure I have labeled the four points $(4, 2)$, $(-4, 2)$, $(-4, -2)$, $(4, -2)$ in the respective first, second, third, and fourth quadrants.

The pair, (x, y), of *numbers x,y* just described is an *ordered pair of numbers* (because the *first* number, *x*, corresponds to distances measured in the fixed direction *XOX'*, the *second*, *y*, to distances measured in a *different* fixed direction *YOY'*). The *ordered pair* of numbers (x, y) is called *the coordinates of the point which this pair represents on the diagram*.

All this detail may have irritated the reader. If so, I can only say, "Go thou and do likewise." Thou wilt have no difficulty in doing for ordinary 'solid' space of *three* dimensions what Descartes did for a plane—space of *two* dimensions. Instead of *two* straight lines intersecting at right angles, imagine *three*, and the three mutually perpendicular planes they determine. But what then? This is only the beginning of a long story whose last chapter has yet to be written. We shall have to consider several mathematically and scientifically important episodes of this story in later chapters, and unless we have understood thoroughly what Descartes did, we shall make but little sense of the work of his successors. For the moment we must notice a few immediate but far-reaching consequences of Descartes' invention.

The *analytic geometry* developed by the use of coordinates is sometimes called *Cartesian* in honor of Descartes.

❖ 7.3 ❖ *Unnecessary Difficulties*

What has become of the point P which is 'represented by' its coordinates (x, y)? What 'is' it? Euclid defined a point as that which has no parts and has no magnitude. This is incomprehensible unless we understand what is meant by 'parts' and 'magnitude,' to say nothing of 'no.' Another familiar definition of a point, as that which has position only, also is obscure. What is 'position'? All these inherited and unnecessary difficulties are swept away by *defining* the *point* P as the ordered number pair (x, y). The mysterious 'point' that eluded Euclid has vanished leaving not even a ghost behind.

This does not imply that the traditional *language* of geometry is no longer useful and suggestive, for it is. The unmystical definition of the 'point' P as (x, y) is merely the first and most readily understood instance of two twentieth-century philosophies, one of mathematics, the other of theoretical physics. Mathematicians long ago gave up trying to define such elemental things as 'point,' 'line,' and so forth as images of 'entities' (that shibboleth of the metaphysicians) 'existing' (another) in some realm of 'ideas' (another) inaccessible to human experience. Instead of seeking inaccessible entities, mathematicians take as the 'elements' out of which 'space' and its 'geometry' are to be constructed certain notions which are undefined beyond the postulates, which the mathematicians prescribe at will, that these notions must satisfy. To two of them they give the names 'point' and 'line,' which might as well be 'joint' and 'jine'; the words are mere blanks in the statements of the postulates. For example, two joints determine a unique jine; two jines determine a unique joint. What is the advantage? Simply this: the 'joints' and 'jines' suggest interpretations other than the 'points' and 'lines' of tradition and visual habit for which the abstracted postulates are valid.

The scientific analogue of this disregard of humanly unknowable 'entities' entered science with twentieth-century physics. In the nineteenth century a physical theory, such as that of light, was acceptable only if it could be represented by a mechanical model. For an extreme instance the reader may consult Lord Kelvin's *Baltimore lectures* on molecular dynamics delivered at the Johns Hopkins University in the late summer of 1884. Probably these lectures were the high-water mark of a great tide which had been rising steadily all through the nineteenth century and which was to ebb out beyond the horizon in the 1930s. The tide was an increasingly complex description of the physical universe by means of imagery in terms of Newtonian mechanics. In his first lecture Kelvin devised "a model molecule consisting of a thin rigid shell to whose interior masses were attached by springs." He believed in his model: "It seems to me that there must be something in this, that is, as a symbol, is certainly not an hypothesis, but a certainty." (The syntax may leave something to be desired, though the meaning is plain.)

As late as the 1920s students of physics were taught to derive Maxwell's[1] equations of the electromagnetic field— at which we shall look later [17.4]—from one of several complicated and abstruse sets of physical assumptions. Since about 1930, at least in advanced instruction, these equations are stated as postulates. Einstein was originally responsible for this reasonable reform. The equations are an adequate mathematical description of what happens in elec-

---

[1] It is customary now to drop the 'Clerk' from Clerk Maxwell, although he was usually called this in the nineteenth century.

tromagnetism under certain experimentally determinable conditions. They are also the simplest descriptions of this sort yet constructed. Would it not then seem to be rather pointless to befog the simple, adequate descriptions symbolized in these equations by seeking to deduce them from artificial hypotheses? There is so much to learn in this twentieth century that it is humanly impossible to retrace all the steps by which we have arrived at our present outlooks. In retrospect some of those tentative and laborious steps appear to have been in wrong directions, and some actually were. The landscape changed as we progressed, even as our perceptions changed.

Again, it was found in the decades following 1925 that much of atomic physics could be compactly summarized and enriched by means of the more modern quantum theory. When this theory first burst on a disconcerted science it was accompanied by the most agonizing appeals to a nonexistent physical intuition that physics had then known. Gradually the mysticism thinned, until many (but not all) workers in this department of mathematical physics were content to accept the basic equations of the theory, from which all the rest is deduced, as postulates. As Dirac, one of the founders of the theory, said in 1930 (*The principles of quantum mechanics*, 1st ed., p. 7), "The only object of theoretical physics is to calculate results that can be compared with experiment." Materializing mathematical abstractions as 'wavicles,' in verbal analogy with the 'particles' of an older dynamics, illuminated no discussion. Nor, on the technical side, did this graphic language add anything but confusion to what, after all, was a matter of pure mathematics. However, if analogies and the like, later discarded as illusory, aided in the construction of a usable physics of atoms, they cannot be dismissed as valueless. If only we knew on starting where we wished to go, we might get there in a step or two instead of wandering miles out of our way. But we seldom know till we have arrived.

It seems not too much to claim that Dirac's down-to-earth view of theoretical physics was a remote but direct consequence of what Descartes made possible in geometry when he supplanted Euclid's 'point' by an ordered pair of numbers, perhaps without sensing what he had done. But it must be noted that in casting out one imaginary devil Descartes made it easy for two very real ones to enter, 'number' and 'order.' What are these? For the present we shall accept them as intuitively clear, although considerably later [Chapters 19, 20] we shall recognize them as the sources of some of the paradoxes and confusions that seem likely to disturb both the Queen of the Sciences and her Servant for many years.

❖ 7.4 ❖ *Three Suggestions*

We must now return to geometry.

The (x, y) diagram immediately suggests two problems.

(1) If a point P traverses a line (straight or curved) in the plane, what is the equation connecting its coordinates (x, y)?

(2) If we are given any equation connecting x,y, what curve does (x, y) trace, and what are its 'geometrical' properties? That is, as x and y *vary*, how does the point whose coordinates are (x, y) move?

(3) If several equations connect x,y, instead of only one as in (2), what is the geometrical equivalent of the connection?

One example for each of these will be enough. If (x, y) is constrained to an ellipse whose center is at $(0, 0)$ and whose semi-axes are a,b the equation of the ellipse is $(x^2/a^2) + (y^2/b^2) = 1$. The proof of this may be left to the reader, with the hint that the coordinates of the foci of the

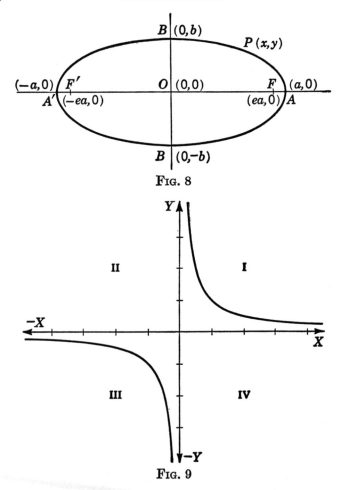

FIG. 8

FIG. 9

ellipse are $(\pm \sqrt{a^2 - b^2},\ 0)$, and one definition of an ellipse
is the curve traced by a point P the sum of whose distances
from F and F' is constant (Figure 8).

For (2), we may plot the curve (a hyperbola) connecting
the pressure x and the volume y of a perfect gas according
to Boyle's law, pressure $\times$ volume $=$ constant (see Figure

9). Without loss of generality the constant may be taken as 1. The equation of the curve is thus $xy = 1$. Here the graph gives more than is physically wanted. What is the gaseous interpretation of the situation represented in the quadrant III, where both pressure and volume are negative? Mathematics habitually confronts us with similar absurdities. Here the escape is obvious: only quadrant I has any physi-

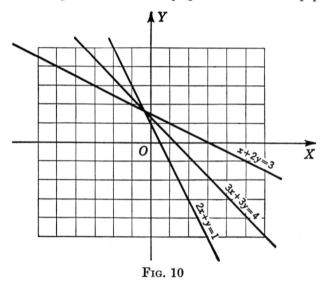

Fig. 10

cal significance. But what about less obvious mathematizations? Such are the happy hunting grounds of speculators on the formal symbolism of free will and determinism as revealed by the mathematics of probability. We shall recur to this in Chapter 18.

To illustrate (3) we refer to school algebra. A straight line is graphed by an equation of degree 1 connecting the coordinates (x, y). Thus, in Figure 10 are depicted the graphs of $2x + y = 1$, $x + 2y = 3$, $3x + 3y = 4$. If all the straight lines represented by these equations meet in a point, the equations will have a common solution $x = a, y = b$, where (a, b) are the coordinates of the point of intersection. Here $a = -\frac{1}{3}, b = \frac{5}{3}$. If the corresponding lines do not intersect, the equations have no common solution.

It is unnecessary for our purpose to pursue such questions. But it may be remarked that the Cartesian revolution in geometry manifested itself most conspicuously in questions of the types (2), (3). Theoretically, at least, algebraists can imagine equations ad infinitum connecting x,y. It is then the task of geometers to translate all these equations into 'geometrical language.' The translation is accomplished a step at a time. In the first stages of this interminable dictionary, curves are classified according to the degrees of the equations connecting the coordinates of any point on them: $x^3y + y = 1$ represents a certain curve 'of the fourth degree'; $x^n + y^n = 1$, where n is a positive integer, represents a curve 'of the nth degree'; and so on. Here there is no limit to the 'geometry' that may be imagined.

The Greeks got but a little way beyond the conics (curves of the second degree). Today a schoolboy after a year of analytic, *Cartesian*, geometry can follow Descartes into the trackless infinities of curves more complicated and incomparably richer than all the lines ever imagined by the most imaginative of the Greek geometers. But this is not the most important advance that analytic geometry made possible to science. The farthest development of it all in this direction is nothing more significant than a technical elaboration of commonplace graphing.

❖ 7.5 ❖ *Intuition into Algebra*

Having seen what Descartes did, we may ask why he should have done it.

Mathematicians do not differ very greatly from other mortals in their ability to perceive space relationships. Only a very small percentage of professional mathematicians are highly gifted in space perception. The majority find a complicated figure, especially in three dimensions, as difficult to see through as anyone whose geometrical training has been negligible.

To take a very simple example, anyone with naturally acute spacial vision will see through the following puzzle instantly. It was invented by A. F. Möbius (1790–1868), whom Gauss considered his most gifted pupil. We shall meet this ingenious man again when we come to topology. A strip

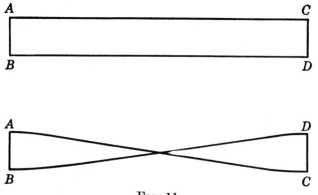

Fig. 11

of paper *ABCD* is laid flat on the table. The end *AB* is held flat while the strip is twisted once so that *CD* is reversed to *DC*. The end *DC* is then pasted onto *AB*, *D* going on *A*, *C* going on *B*. How many sides has the twisted band formed by the pasted strip? Suppose next that in the original strip (before twisting) *CD* is pasted onto *AB*, *C* on *A*, *D* on *B*. The result is a cylinder with an inside and an outside. This cylindrical surface has two sides. *The twisted strip has only one side.* Yet it is a safe wager that nine persons out of ten would assert that a surface with only one side is an impossibility—until they had seen it constructed. Those who guessed right may now amuse themselves by imagining how an electric charge would distribute itself over a one-sided conductor.

Three further puzzles, easier to dispose of than the preceding, may suggest that spacial intuition is not a universal gift. How would you cut a single slice off a cube, the base of the slice to be a flat surface, so that the sliced part of the cube shall be bounded by a regular hexagon (six equal sides and six equal angles)? If this is too easy, an exercise from the first six months of school geometry will illustrate another kind of cut-and-try. In Figure 12 A', B', C' are the

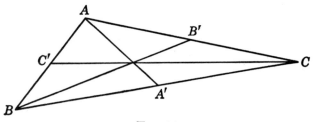

Fig. 12

midpoints of the sides of the triangle ABC. Prove that the straight lines AA', BB', CC' meet in a point. Finally, as a harder exercise, prove the following theorem discovered by G. Desargues (1593–1662). (I state the theorem as it is usually given in college geometries. Certain refinements to take account of the full projective plane, of which Desargues knew nothing, would be necessary to include such exceptional cases as that in which sides of the two triangles are parallel.) ABC, XYZ are triangles *in the same plane.* The three pairs of straight lines (prolonged, if necessary) AB and XY, BC and YZ, CA and ZX meet respectively in the points U, V, W. It is to be proved that U, V, W lie on a straight line. If the triangles are in *different* planes not parallel to each other the proof 'leaps to the eyes.' Can you prove the theorem, as stated, using only Euclidean plane geometry?

Probably most readers of this book have solved the second problem at some stage of their careers. The majority of those who did will recall that some disreptuable trick had to be devised by hit-or-miss before the proof could be constructed. There was *no infallible method* for attacking such a problem which anyone who had mastered the technique could apply and be reasonably sure of success. Nor had the man who first discovered this proposition, perhaps by accident from crudely drawn diagrams, any method for testing the truth of his conjecture that AA', BB', CC' really do meet in a point. It is easy to draw pictures of false propositions that even the most refined drawing-board constructions suggest are true. What is wanted here is some automatic machinery for winnowing the true (logically consistent) from the false. Descartes invented such in his analytic geometry.

The power of Descartes' method is twofold. First, it reduces the proof or disproof of any conjectured proposition to an algebraic routine which any intelligent boy of sixteen can learn to apply more or less mechanically whether gifted with space intuition or not. Second, and more important, the mere language of geometry when thus translated into algebra becomes infinitely more suggestive and in fact does frequently indicate unsuspected geometrical relationships. When fully developed, as it is today in a college course, the method of Descartes showers us with unsought geometrical theorems that even the most penetrating space perception—what the Greeks relied on—might stare itself blind trying to see.

If the ability to penetrate a disordered cobweb of lines and surfaces to the hidden simplicity of a well-concealed geometrical theorem is rare, the gift of being able to visualize complicated motions of actual bodies is even rarer. Here another 'dimension' is imposed on the three (length, breadth, depth) of solids, namely, *time*. Instead of static rigid bodies with only three dimensions we now have dynamics—in its earliest stage the description of the motion of a single particle acted upon by forces—a geometry of *four* dimensions.

In passing it must be emphasized that this is not equivalent to the nonsensical statement that 'time is the fourth dimension.' It will be seen later in what sense time is *a* 'fourth dimension' in the abstract language of mechanics. Time is not *the* fourth dimension, for the sufficient reason that '*the* fourth dimension' is a collection of three words without meaning. It may be noted incidentally that all the apparent paradoxes of *the* fourth dimension in scientific fiction and misapplications of geometry to eschatology come from insisting that the irreducible elements of a space of four dimensions must be *points*. That a point-space of any number of dimensions is not the only one thinkable, will appear as we proceed.

As for the difficulty of visualizing the phenomena of motion intuitively, it is sufficient to think of a spinning top, a gyroscope, or a baseball pitched with an incurve. Mount a gyroscope in a box, set the gyroscope in rapid spin, and predict what will happen when the box is struck. A genius at space-time intuition—merely a more subtle form of ordinary spacial perception—should be able to foresee what must happen before the gyroscope is put in the box. He should also be able to predict what will happen when one smoke ring overtakes another. Try it and see.

By making a geometry out of dynamics the followers of Descartes reduced the science of motion to the same attainable level as analytic geometry. They eliminated the necessity for repeated intuitions.

This does not imply that intuition in mechanics and other parts of science is useless or not practiced. Intuition is still as necessary as it ever was. The all-important first step of translating phenomena into mathematical symbolism is still almost wholly intuitive, and the fruitfulness of a particular theory is largely predetermined by the instinctive scientific vision of the man who first sees what is happening. But once this step has been taken it is now no longer necessary to repeat the heartbreaking struggle, as in scaling Mt. Everest—before it was flown over and the problem of climbing it was reduced to triviality—before taking the next step. Mathematics is the guide, and progress into an untraveled wilderness becomes an automatic process of putting one foot after the other.

❖ 7.6 ❖ *Algebra into Geometry*

Although but few are gifted with acute spacial or motile intuition, all of us have the little we acquired when we learned to walk. To a slight degree it is easy for us to think in terms of geometrical forms. Such thinking seems to come naturally to us, or we imagine that it does; whereas thinking in the symbols of algebra and analysis is a habit that must be acquired deliberately. Without our ability to manipulate abstract symbols, our attack on the complexity of nature would be hopeless. It is at the infantile level of abstraction that crude geometrical intuition colors the language of science and determines its imagery.

The second great thing that Descartes' invention of analytic geometry introduced was the simple vocabulary for talking geometrically about situations which are not spacial in the sense of our childhood intuitions, but algebraic or analytic in the sense of highly sophisticated mathematics. Thus the 'natural' language of geometry when applied to analysis suggests useful things to do in the abstract analysis itself.

A space of four or more dimensions is not an intuitive
notion. Yet such a space becomes as simple as a familiar
room when we transfer to the elementary algebra of systems
of equations (and functions) in *four* variables the *language*
which for *three* variables describes relations between points,
lines, surfaces, volumes, distances, and angles. Nor is there
any mystery about the metaphorical extension of 'distance'
or 'curvature' to a space of any number of dimensions. It
is all a convenient manner of talking, and to seek anything
more abstruse in any of it is to look for magic in empty
words. In science only misguided and deluded mystics like
the astronomer and physicist J. K. F. Zöllner (1834–1882),
of spiritualistic fame, believe that mathematical manipula-
tions performed on sets of four independent variables call
into existence a fourth dimension of space in which they will
some day be reunited with their defunct grandmothers.
When millions or billions or an infinity of dimensions are
conjured onto the paper by a stroke of the pen, even the
most hopeful begin to despair of ever locating their ancestors again. We shall recur to this in the next chapter.

So far as its utility in science is concerned, the metaphysics of 'space' can be ignored. It is never necessary to
inquire what 'space' is or whether it 'exists,' or even to
decide whether either of these alluring questions means
anything at all. We need only know how to manipulate
equations whose variables denote numerical measures occurring in the laboratory or in common experience.

In simplifying geometry and extending its usefulness,
Descartes practically abolished 'space' and terminated a
debate which had raged for centuries. It is amusing to
thumb through the philosophical literature of the fairly
recent past, say since the seventeenth century, to observe
how often the 'space' of the philosophers has affected sci-

ence. With the unflattering exception of Kant's erroneous ideas, obliterated by the invention of non-Euclidean geometry in the 1830s, there seems to be no instance worth noting. I am aware that Kantian apologists say that Kant never said what he explicitly said, and assert that his long-outmoded conceptions of 'number' and 'space' accommodate all the developments of a mathematics and a science of which he had no glimmer of foresight. There is one distinction between some philosophers and most scientists that may be significant. When a scientist makes a mistake, he either admits his error and strives to correct it, or his colleagues and pupils see that he does.

Yet although the metaphysical problems of space and time so far have been irrelevant for the kind of reasoning that has proved usable and fruitful in science, it may not always be so. But it would seem that since the seventeenth century the metaphysics of physics is a luxury that only a few creative physicists can safely afford. Against this somewhat narrow view of what science is about, even at its most technical, must be placed the fact that a few revolutionary scientists, like E. Mach (1838–1916) and Einstein, have been motivated by impulses that were philosophical rather than strictly scientific. At least that is how some of the decisive innovations of twentieth-century physical science strike a mere outsider—"In my Father's house are many mansions."

Against the tremendous gains that science acquired with Descartes' invention must be put a possible loss. How do we know that we are on a reasonable track at all, and are not blindly following our childish bent for trivialities because ingrained habit has made them as familiar to us as our own hands and feet? Manipulating our mathematical descriptions of nature, our hands fall into the accustomed tricks which rude experience has taught us. The table was a very rigid body when we bruised our ankles on its square legs. Carrying this experience and countless others—visual and tactual—with us all our lives, we naturally 'geometrize' our analysis. It does not follow that the things we make our equations tell us in the one-syllable words of elementary geometry are all that they might tell or even the most suggestive things that might be told. By forcing a geometrical vocabulary on analysis, and hence also on our abstractions of nature, we may be compelling the universe to speak an artificially simple language. When we come to 'abstract space' in the sequel we shall see instances of this kind of language.

We have not yet outgrown the Greek way of phrasing everything scientific in the primitive language of geometry. The Greeks had no analysis. Possibly we shall outgrow their language and begin to talk something closer to our own minds when elementary geometry yields precedence, as it should have done long ago, to algebra in school mathematics.

If the Greeks had possessed anything as powerful as our algebra and the calculus [Chapters 14, 15] they surely would not have restricted themselves to the limitations of pictorial thinking. That they did is not a valid reason for mathematicians teaching generation after generation of future scientists to do likewise. Real mischief is done when the credulous pupils acquire an ineradicable belief that their purely metaphorical language describes either an 'existent space' or an 'objective reality,' whatever, if anything, those high abstractions may mean for a consistent science. "God ever geometrizes," said Plato, and he said it for two reasons: he knew next to nothing about geometry, and neither algebra nor mathematical analysis had been invented in his day.

Chapter 8

LANDMARKS OLD AND NEW

❖ 8.1 ❖ What Is Geometry?

A more appropriate question for our immediate purpose would be 'What *was* geometry in its second golden age of the nineteenth and early twentieth centuries?' By the 1920s geometry had already entered a new phase, vaster and more powerful than ever, partly because the general theory of relativity (1915–1916) had suddenly revealed new horizons. But relativity was by no means the whole story. Geometry gradually became abstract. Of this more later [8.9]. The newer geometry goes far beyond that which I am about to describe, vast as that was, and it is of unprecedented importance for its suggestiveness in the physical sciences.

The spirit of geometry from 1872 to about 1922 cannot be better or more briefly described than in a famous sentence (1872) of F. Klein. All the astounding inventiveness and infinite variety of geometry during that amazingly prolific half century is seen as one orderly, simple whole from the commanding summit which Klein recognized as the proper point of view to sweep in nearly the whole of the past of geometry as it was in his day and to foresee much of its future. Here is the famous sentence: "*Given a manifold and a group of transformations of the same, to develop the theory of invariants relating to that group.*"

It is a pity to spoil the beautiful simplicity of this by explanations, but I can be brief. *A manifold of n dimensions* is a class of objects which is such that a particular object in

141

the class is completely specified when each of n things is given. For instance, a plane is a *two-dimensional manifold of points*, because the plane can be considered as the class of all its points and any point in the plane is completely specified, or uniquely known, when its *two* coordinates x and y are given [7.2]. Common solid space similarly is a *three-dimensional manifold of points*. For the moment I shall leave it to the reader to see that common solid space is also a *four-dimensional manifold of straight lines*. This should rob '*the* fourth dimension' [7.6] of some of its mystery. If this is not sufficient the next section, on space of many dimensions, will offer anyone who may be interested sufficient hints for the construction, on a drawing board, of space of any finite number of dimensions. Later we shall come to Riemann's manifolds, essentially those described above, and see their scientific importance.

The *transformations* referred to are of the kind which replace each object of the manifold by some definite object of the manifold, or even of another manifold. For instance, we might consider all those transformations of the straight lines of solid space which carry straight lines into other straight lines, or into spheres, for (as the reader may easily think out for himself) common solid space is *four* dimensional in spheres as well as in lines. It takes *four* numbers to fix a particular sphere; three to fix the coordinates of the center, and one to fix the length of the radius. *The number of dimensions of any space depends only upon the elements (points, lines, planes, spheres, circles, and so on) in terms of which the space is described.* This was the kernel of Plücker's theory of geometric dimensionality which will be described in more detail shortly.

The transformations, according to Klein, must form a group. The postulates for a group are given in the next chapter, and these postulates are the official definition of a group. But as the group is the central and commanding concept of Klein's whole vast program, let us look at its leading property, that on which Klein relied and to which he had reference, although he did not understand the term 'group' as it is used today.

Consider a class of things and a set of operations which can be performed on the members of that class. If the result of performing any one of the operations upon any given member (or members) of the class is a uniquely determined member of the class, we say that the class has the *group property* with respect to the operations. The class then is *closed* under the operations of the set. Thus the class of positive whole numbers, 1, 2, 3, . . . , has the *group property* with respect to addition, for the sum of any two of these numbers is again one of the class. The like holds also for multiplication, but not for subtraction or division.

The *invariants* in Klein's program are those things (properties, actual figures, or what not) that persist, or remain unchanged, under all the transformations, or operations, of a particular given group of transformations, such as translations, reflections, projections [6.1], and (vaguely) the like —all, however, *linear* (of the *first* degree in their variables).

Finally, notice that nothing is said about the number of dimensions of the manifold. This may be 1, or 2, or 3, . . . , or *n, or it may be infinite.* All possibilities are envisaged in the vast program.

Was Klein's program simply an empty dream, an unnecessary abstraction and generalization of the familiar? Far from it. In its concreteness and applicability to the geometry of its epoch it differed from some of its successors and rivals. From Klein's comprehensive point of view the geometers of the second golden age of geometry saw projective geometry, metric geometry of all kinds, Euclid's geometry, innumerable non-Euclidean geometries, geometries of any number of dimensions [7.1], and much more, as harmonious parts of one comprehensive, simple program. It was one of the memorable things of all mathematical history, not merely an outstanding achievement of the nineteenth century. That the present has gone far beyond Klein, and has ascended higher than he saw, does not diminish the sublimity of his conception. After all, we may reflect that if Klein is dead so is Shakespeare.

❖ 8.2 ❖ *Further Progress*

As it is impossible, at least in mathematics, to say everything at once, I shall presently anticipate once more from Chapter 9 a detail about groups as the term is now commonly understood. Klein's reliance on the group property of closure of a class under specified operations, just described, led him to his comprehensive unification of the geometries of his day. But in the 1870s when Klein proposed his comprehensive synthesis, and indeed until the early 1900s, no precise and generally accepted definition of a group as the term is now accepted had been formulated. Consequently many statements about groups and their applications in the older classics of mathematics, including Klein's, must be read with close attention to see exactly what they mean for mathematics as it evolved after 1900. The relevant detail here in all this is that Klein's profound unification includes one of the most extensive kinds of geometry of the twentieth century, topology, only by stretching courtesy.

The anticipated detail from the current and universally accepted definition of a *group* is this. For each of the operations in a *group* of operations there is another operation, uniquely determined, such that the result of performing the two operations successively in either order leaves whatever is operated upon unchanged. Each of the operations concerned is called the *inverse* of the other. For example, if one operation is the shifting of a segment of a straight line one inch along its length to the right, the inverse operation is a shift along one inch to the left.

Klein and his contemporaries, also his immediate successors, sometimes ignored the existence of inverses for the operations with which they were concerned, and at other times they tacitly assumed that the operations had inverses when they had not. The sharpening of the earlier work, including Klein's, ultimately relaxed the definition of a 'geometry' to the study of a 'geometrical object'—anything a professional geometer might be interested in—as an *invariant* with respect to a prescribed closed set of transformations. This of course was not intended as a final description of what geometry is about, and it probably will be superseded, if indeed it has not already gone the way of all attempts to hobble geometry in a sentence. As Klein in looking back over his long career remarked, it is the inevitable lot of mathematical works, even masterpieces, and their authors to be superseded, and this is no tragedy but an inescapable concomitant of progress as opposed to stagnation. Klein himself shortly before his death in 1929 acknowledged that his successors had far outstripped him.

❖ 8.3 ❖ *Space of Many Dimensions*

We have noted [3.3, 7.1] what Descartes' invention of coordinates suggested for a geometry of n-dimensional space, where n is any positive integer, and also, going far beyond this, for space of an infinity of dimensions. But there is another road among many into the outskirts of the unbounded territory of polydimensional space, and this particular way has some advantages that others lack.

The specific idea of dimensionality (this is not the most

general) to be described was conceived by J. Plücker (1801–1868). Plücker started as a geometer, but for many years abandoned mathematics for physics, where he is still remembered, because his mathematical colleagues were too beetle-eyed to see the vast scope of his comprehensive new outlook on 'space.' At the time they were bedazzled by a spectacular resurrection of the synthetic projective geometry of the seventeenth century. Cayley in 1843 and the universal genius H. G. Grassmann (1809–1877) in 1844 independently reinvented Plücker's theory (1831) of dimensionality, too late to do the genial but discouraged Plücker much personal good. The way of the original transgressor in mathematics is sometimes as hard as it is in respectable society. Fashion as king is sometimes a very stupid ruler.

As was observed a little way back, the kernel of Plücker's theory of geometric dimensionality is that the dimensionality of a given space is not an absolute constant, but depends upon the elements, accepted as irreducible, in terms of which the space is described. Descartes might have said that a plane is a space of *two* dimensions because precisely *two numbers,* or *coordinates,* are necessary and sufficient to locate any particular *point* in the plane with respect to a pair of intersecting straight lines, *the axes of coordinates* [7.2]. Similarly it is seen, but seldom noticed as it might be, in an elementary course in plane analytic geometry, that precisely *two* numbers are necessary and sufficient to identify any particular *straight line* in the plane—these numbers are *the coordinates of the line.* So we say that a plane is *two-dimensional in both lines and points when either are taken as the irreducible elements out of which the plane is composed and from which plane geometry is to be constructed.* Nothing, except possibly some peculiar quirk in the rods and cones of our eyes, compels us to visualize space as a swarm of coplanar points instead of as a flat haystack of lines.

If instead of either points or lines as the irreducible elements of a plane geometry we choose circles, the plane is *three*-dimensional. For it takes precisely three numbers to identify a particular circle in the plane, two for the coordinates of the center of the circle and one for its radius. As another example, it requires precisely five numbers to identify a particular conic section (curve of the second degree) in a plane; the plane is *five*-dimensional in conics. For those who remember some analytic geometry it should be clear that the set of all conics in a plane passing through a fixed point is a *four*-dimensional manifold.

I have already alluded [7.5] to the dynamics of a particle as a four-dimensional space of points—three coordinates for the position of the particle and one for time. If instead of points for ordinary space we take straight lines as elements, the space is four-dimensional. This is so because in ordinary solid analytic geometry precisely four numbers are required to specify a particular line. The resulting *line geometry* finds an immediate application in the kinematics of rigid bodies in ordinary three-dimensional space. It was extensively developed for the convenience of those who think more easily in spatial imagery than in algebraic symbolism. The mere language of geometry, ingrained by centuries of use, suggests profitable things to do in this kinematical geometry as in other geometries of more than three dimensions.

❖ *8.4* ❖ *Duality*

Note in the preceding section the emphasis on the *two*-dimensionality of a plane in *points* and in *lines*. This is the historical germ of the *principle of duality* in plane projective geometry, whereby from a statement concerning points (or lines) a *dual* statement concerning lines (or points) can be immediately inferred without independent proof.

For example, two *points* determine a *line,* the *join* of the two points; two *lines* determine a *point,* the *intersection* of the two lines. As it is usually first presented, plane analytic geometry describes a curve as a *locus of points.* In the dual description, a curve is *an envelope of lines;* that is, it is visualized as being cased in by the lines tangent to the point-locus. Incidentally, this may first have been observed by grain merchants exhibiting samples of their wheat. At county fairs years ago in England the merchants would scatter a few handfuls of wheat on a counter and then sweep out beautifully clean curves in the grains with a long straight rod.

Thus by interchanging the words 'point' and 'line,' 'join' and 'intersection,' with some other equally simple devices suggested by geometrical intuition and common logic, plane projective geometry [6.1] is doubled at one stroke. 'Point' and 'line' are duals, as are 'join' and 'intersection.' The duality was extended to metric geometry, but not intuitively. In ordinary space of three dimensions 'point' and 'plane' are duals, 'line' is self-dual—a line corresponds to itself, or is left invariant, in the dualistic interchange of 'point,' 'plane.'

Plücker generalized all this to a *principle of duality* for any two classes of configurations having equal dimensionalities and being of the first degree in their respective coordinates, equal in number to the common dimensionality. Once the esoteric knowledge of a few, all Plücker's innovations long since became commonplaces to the many—"For all have got the seed now, and most can grow the flower." But the accident of popularity is not necessarily a cheapening of a great idea or a degradation of high art.

Finally, in the 1920s, an extensive theory of dimensionality in the spirit of modern abstract mathematics expanded rapidly, and almost as rapidly in the 1930s and 1940s contracted to reasonable proportions.

❖ 8.5 ❖ Non-metric versus Metric

Topology has been mentioned in passing. We must now see very sketchily how topology as a species of geometry differs from both the common Euclidean geometry of the elementary high-school course and the classical projective geometry, already noted [6.1], of a college course. It is also far different from the extension of Descartes' geometry to that of a space of more than three dimensions.

A root of the fundamental distinction between topology and other kinds of geometry is in the kinds of permissible transformations. To see the difference, I recall what we were allowed to do in school geometry. It was assumed there that triangles and other figures could be slid about in a plane without alteration of measures of distance—lengths of sides—and sizes of angles. It was also sometimes tacitly assumed that one triangle could be lifted out of a plane and then be superimposed on another without having undergone distortions of measurements in the process. Neither assumption is justified without explicit postulation, as there are 'spaces' and 'geometries' in which at least one is not permissible. Euclid, evidently relying on what to him seemed obvious, omitted both assumptions, among several others, equally 'obvious,' from his list of postulates. But in projective geometry, as we noted [6.1], the lengths of the sides of the figures projected into others had no relevance for the theorems sought and exhibited. As a simple but sufficient example, if two lines in the original figure intersect they continue to intersect in the projection. (The sophisticated reader will make allowances for the region at infinity and the full projective plane. The unsophisticated reader will get the essential sense of what has been said from his visual experience. This is enough for an account like the present.) In central projection, for instance, the projected figures are the shadows of others thrown on a screen by a point-source of light. Among the permissible transformations in Euclidean geometry are those allowing rigid bodies to be moved about freely in space without distortion. In projective geometry when phrased algebraically the permissible transformations are linear (of the first degree) in the coordinates. In topology the transformations are not restricted to be even algebraic, and it is required to find what remains invariant under certain very general types of transformations. These will be accurately described later [8.8].

Topology originated in what now would be called mathematical recreations. I shall recall only two of several. It seems that in the eighteenth century two islands in the Pregel were connected by a bridge and with Königsberg by six bridges, and some disturbing soul asked whether it was possible to start from the mainland and traverse each of the seven bridges, each one *only once*, and return to the mainland. As a matter of fact it was impossible, but this required proof. Clearly it would not affect the problem if the island and the mainland were swelled or shrunk without touching, and if the seven bridges were twisted in any way so long only as no two of them were permitted to intersect. Euler (1736) disposed of this puzzle.

The second recreation, so-called, goes very much deeper. A solution was still lacking in 1950. Practical cartographers had noticed that a map of any plane area, say that of a continent cut up into countries, could be colored in at most four colors. In such a map the same color may be used for each of several different countries provided no two touching along a line boundary are colored alike. The ocean, of course, is counted as just another country, surrounding all the others if the land mapped is an island. The reader may easily construct a map which requires exactly four colors. The problem is to prove that, no matter how complicated the map, four colors suffice. The problem was noticed as early as 1840 by Möbius, and again about 1850 by A. De Morgan, a founder with Boole of modern logic, a born non-conformist always on the lookout for something unusual, and, incidentally, the father of the ceramic artist and novelist (*Alice for Short*) W. F. De Morgan (1839–1917). Cayley (1878) advertised the problem to professional mathematicians and commended it to their attention. He evidently had spent some time on it. Several attempts to prove the sufficiency of only four colors for a plane map followed Cayley's appeal for a rigorous solution. Some seemed promising, but all were incomplete or fallacious. It makes no difference to the problem, as in that of the Königsberg bridges, whether the map to be colored is deformed in any way that introduces no new boundaries. Nor does it matter whether the map is drawn on a plane or on a sphere. But as the reader will see after reading [8.6], the map problem on a doughnut differs from that on a plane—the connectivities are not the same for a plane and a doughnut. The problem is thus one of *topology, that division of modern geometry which deals with the properties of figures unchanged by continuous* [4.2] *deformations*. This will be made more precise later [8.8]. The map problem belongs to what is called *combinatorial topology*, in which *continuity*, as for functions of a real variable [4.2, 6.2], is of comparatively minor importance.

My topological friends assure me at considerable length that combinatorial topology is essentially worked out, or about to be so. It is partly an abstract and extremely complicated but approachable problem in the technical theory of groups [Chapter 9]. But my topological friends do not say why combinatorial topology had not as late as 1950 disposed of the map problem, then over a hundred years old. Of course it may do so the day after tomorrow. Coloring a map in the most economical way may after all be of no interest to an expert topologist of the twentieth century. The situation here is singularly like another I shall describe [11.4], in the theory of numbers, where Fermat's famous 'Last Theorem' of about 1637 is said to be of but slight interest to experts in the complicated theories devised without success to dissipate the original difficulty responsible for the elaborate theories. Mathematicians after all may be as human as the rest of us, perhaps even more so.

Definite progress in the four-color problem is disappointing. It has been limited to proving that solutions exist for a map with a given rather small number of regions. The labored record up to 1940 does not seem to have been beaten; for any map of thirty-five or fewer regions four colors suffice. The problem, like Fermat's, is one of those easily stated and deceptively simple things that amateurs had better leave alone. The professionals seem to have given it up. As a personal reminiscence I recall that G. D. Birkhoff said shortly before his death that in spite of all his efforts, one of which I witnessed in 1911, to crack the four-color problem wide open he had not even scratched it. Still, some uninhibited explorer breaking a new trail may reach the end tomorrow. It may be that sometimes in mathematics too much knowledge is a shackle.

There is another very famous topological problem that at a first look seems as hard as the map problem. In fact Poincaré abandoned it, and shortly before his death (1912) proposed it to all mathematicians as a question worth their consideration. For about a year it was called 'Poincaré's Last Theorem,' in analogy with Fermat's, because it was anticipated that whatever Poincaré could not settle must be really hard and likely to remain a challenge for many years. However, in 1913 Poincaré's conjecture—it was only that—was proved correct by G. D. Birkhoff. It is given that a continuous one-one transformation (compare [8.8]) takes the ring bounded by two concentric circles into itself in such a way as to advance the points of the outer circle positively and those of the inner circle negatively, and at the same time to preserve areas. It is to be proved that there are at least two points invariant (left fixed) under this transformation. This may seem like a useless puzzle. Actually Poincaré had reduced a difficult problem in dynamical astronomy—the restricted problem of three bodies—to this one in topology.

❖ 8.6 ❖ *Connectedness*

The first fairly general theorem in topology was stated explicitly by Euler in 1752. Descartes, however, had used it as early as 1640. It is often given in the school texts on elementary solid geometry in the form

$$E + 2 = F + V,$$

where, for 'any' polyhedron, E, F, V are the respective numbers of edges, faces, and vertices. The 'any' is much too inclusive. This needs some amplification, particularly by the concept of *connectedness* for a region or surface.

To see intuitively what is meant by connectedness, imagine two circular disks, one with no holes in it, the other with holes. The first has a single boundary, the circumference of the disk; the second has in addition to this boundary the boundaries of the several holes. On the first disk any closed simple curve, say a circle, can be shrunk to a point without passing out of the region delimited by the boundary of the disk. On the second disk this is not always possible, as the curve might surround one of the holes and could shrink no farther than the boundary of the hole and remain on the surface. To shrink to a point, the curve would have to pass into the hole and therefore out of the region limited by the boundaries of the holes. The first disk is said to be *simply connected;* the second, *multiply connected.*

A region or surface S is *connected* if any two points on it can be joined by a continuous arc lying entirely on S. If S is connected, and if every closed contour C on S separates S into two connected parts, of which one has C as its complete boundary, S is *simply connected.* A cross cut joining two points on the rim of a simply connected surface S severs S into two simply connected surfaces.

This digression on connectedness was necessary because the formula of Descartes and Euler applies to any map drawn on a sphere if all the regions of the map are simply connected. The formula can be written

$$V - E + F = 2,$$

where V is the number of vertices, E the number of edges or sides, and F the number of faces or regions. For example, in school geometry the formula applies to a cube blown up, like one of those checkered rubber balls that beautify our bathing beaches, to the surface of its circumscribing sphere, for which $V = 8, E = 12, F = 6$. In this uninhibited geometry the theorem holds for any polyhedron whose faces are simply connected. So we need not have blown up the cube. For a tetrahedron, $V = 4, E = 6, F = 4$. The requirement sometimes prescribed in the school texts that the polyhedron be convex is unnecessary.

❖ 8.7 ❖ *Knots*

Another historic topological problem is that of classifying and enumerating knots. This goes back at least as far as Gauss, who considered it on several occasions, the earliest being 1794 when he was seventeen, and the last 1849, within six years of his death. Gauss prophesied that topology would become one of the major divisions of mathematics. He was right. He did not, however, prophesy that topology would be of interest to his successors of A.D. 10,000 as some of his followers insist.

Though there has been much interesting work on knots, the general problem (at this writing) is unsolved. The best up to 1950 was by J. W. Alexander (1888–), who in 1927–28 defined certain invariants for distinguishing one knot from another. The problem was of some potential scientific interest in the 1870s when the vortex theory of atoms had a deservedly short popularity. Vortex atoms were just too fantastic to make physical sense. Of more lasting scientific interest was the work (1833) of Gauss in electrostatics, where a problem equivalent to the mathematics of both unlinked and interlinked circuits demanded for its solution topological considerations like those in the theory of knots. The somewhat similar investigations of G. R. Kirchhoff (1824–1887) are still part of a regular college course in electricity.

So the problems of knots after all were more than mere puzzles. The like is frequent in mathematics, partly because mathematicians have sometimes rather perversely reformulated serious problems as seemingly trivial puzzles abstractly identical with the difficult problems they hoped but failed to solve. This low trick has decoyed timid outsiders who might have been scared off by the real thing, and many deluded amateurs have made substantial contributions to mathematics without suspecting what they were doing. An example is T. P. Kirkman's (1806–1895) puzzle of the fifteen schoolgirls (1850) given in books on mathematical recreations.

❖ 8.8 ❖ *One Kind of Topology*

Modern noncombinatorial topology is a creation of the twentieth century. In E. Kasner's (1878–) apt phrase, topology is rubber-sheet geometry. Imagine a tangle of curves drawn on a sheet of rubber. What properties of the curves remain unchanged as the sheet is stretched and twisted and crumpled in any way without tearing? Or what are the *qualitative* properties of the tangle as distinguished from its *metrical properties*—those depending upon measurements of distances and angles? To attack this, topologists concentrate on what is happening in the neighborhood of a point as the sheet is deformed. To curtail a very long story, I shall merely state the cardinal definitions. These are all in the modern manner and are necessarily abstract.

A 'space' is a set of '*objects*' together with a set of subsets, called *neighborhoods*, of the original set, and every neighborhood of an object contains that object. Say the space is S, and A is any subset of S. An object is called an *inner* object of A whenever there is a neighborhood of the object that is contained in A. The set remaining when all the objects of A are removed from S is written $S - A$. An object is called a boundary object of A if every neighborhood of the object includes objects of both A and $S - A$. The boundary of A is the set of the boundary objects of A; and A is said to be *open* when its boundary is contained in $S - A$, *closed* when its boundary is contained in A.

So much for neighborhoods. Transformations must now be described. A *transformation* of one space S into another space S' is the assignment of a correspondence between the objects in S, S' such that to every object in S there corresponds at least one object in S' (as in mapping, say); the transformation of a subset A of S is the set of all correspondents, under the transformation, of all objects in A. The transformation is said to be *uniform*, or *single-valued*, whenever it assigns a unique correspondent to every object in S; it is said to be *continuous* whenever the transform (image in the map) of every open set A of S is an open set of S'; and last, the transformation is said to be *homomorphic* when it is one-one and continuous both ways.

Putting all this into a concise statement we have this definition: *Topology is the study of those properties of spaces that are invariant under homomorphic transformations.*

I shall not describe *combinatorial topology*, for the reason indicated earlier [8.5]. The kind just defined is sometimes called *analytic topology*, because of the continuity [4.2] postulated in its permissible transformations. It will be noticed that some of the concepts implied in the final definition occurred in our description of the algebra of logic [5.2].

To conclude these remarks on one of the most active divisions of twentieth-century mathematics, I recall that topology became of first-rate scientific importance in the dynamical researches of Poincaré, particularly [8.5] in connection with the problem of three bodies attracting one another in space according to the Newtonian law of gravitation—for example, the sun and two of its planets. It was a question of describing the families of possible orbits. Numerical calculation was too laborious and too slow to reveal the extremely intricate motions for more than a step at a time. A qualitative attack was indicated, and for this Poincaré created (1895, 1899, 1900, 1904) a major division of topology. He originated a rigorous combinatorial topology for space of any finite number of dimensions. Some of what he did has still to be surpassed. It will be interesting to see what modern calculating machines can do with a problem as complex as that of the motion of the moon. So far—and these were only the great-grandfathers of the post–World War II giants—the machines in three months checked what it had taken one of the most expert computers in all the long history of dynamical astronomy nearly forty years to accomplish. He had made no slip.

❖ 8.9 ❖ *Abstraction Again*

Geometry did not escape the passion of abstraction, sometimes quite furious, that changed twentieth-century algebra into something so new and strange that the algebraists of the preceding century would hardly have recognized it as their progeny. In geometry the process of abstraction followed the same general pattern as in algebra. Underlying concepts were reformulated in abstract terms. In the process the traditional and familiar connotations had been painstakingly extracted and sublimed. In some ways it was like stewing all the juice out of a chicken. The objective was similar to that in algebra, to lay bare the hidden bones of at least some of the swollen theories of geometry.

One outcome was a theory of abstract space; another was the application of this excessively refined theory to a generalization of much of classical analysis [6.2]—the calculus [Chapters 14, 15] and its many offshoots—in what was called abstract or general analysis. All of these abstractions quickly expanded to enormous size if nothing else. Several are still expanding as this is written and doubtless will still be expanding a generation hence. If the original purpose was to simplify and coordinate the vast masses of geometry and analysis inherited from the nineteenth and early twentieth centuries, abstraction may have frustrated itself, like Aesop's bullfrog. The new shortly became even more puffed up than its traditional rivals which it had sought to outswell without bursting itself. Such an ambition, however, is not necessarily discouraging. The cure for mere bigness is more of the same till the inevitable blowup. Then the survivors of the disaster may salvage whatever seems to them worth saving. Only a mere indication of the origin of two concepts of the theory, those of a generalized 'absolute value' and a generalized 'distance' can be described here.

In the algebra of ordinary complex numbers [4.2, 5.3], the *absolute value,* written $|x + y \sqrt{-1}|$, of the complex number $x + y \sqrt{-1}$, where x,y are real numbers, is the real number $\sqrt{x^2 + y^2}$, the positive value of the square root being taken. For example, the absolute value of $3 + 4 \sqrt{-1}$ is 5. In the graphical representation [4.2] of complex numbers the absolute value of $x + y \sqrt{-1}$ is the length of the line *OP* joining the origin $(0, 0)$ to the point (x, y) (see Figure 6).

These absolute values have two basic properties. The first is merely Euclid's theorem that the length of any side of a plane triangle is equal to or less than the sum of the lengths of the other two sides; Euclid omitted the 'equal to,' which refers to degenerate triangles whose three vertices are collinear. This is the *triangle,* or *triangular, inequality.*

The second basic property of absolute values is that the absolute value of the product of two complex numbers is equal to the product of their absolute values, and it is interesting to see that essentially this theorem was known to Diophantus of Alexandria probably in the first century A.D.—certainly not later than the third century. Diophantus of course knew nothing of complex numbers, so he effectively stated the theorem as "the product of two sums each of two squares is a sum of two squares." This generalizes to four squares and to eight squares, but to no other number of squares. The negative result is not easy to prove. Cayley (1881) gave the first proof. The theorem for two squares implies much of trigonometry; that for four squares, some details of the algebra of quaternions [5.8]; and that for eight squares turns up in Cayley's non-commutative, non-associative algebra [5.8]. All of which may suggest further generalizations of the next.

In an abstract commutative field F [3.1] with elements $0'$ (the zero element of F, accented to distinguish it from the zero element 0 of the real number field), $x, y, \ldots, x^2 + y^2$ are not numbers. So if the formal properties of absolute values for complex numbers are to be preserved, the defini- tion of absolute value must be revised at the beginning. The problem was brilliantly solved by what some mathemati- cians of an older generation considered a disreputable trick when J. Kürschák (1864–1933) concocted his definition (1913) for *absolute values* in F. He 'associated with' any element z of F a unique real number, its 'absolute value,' denoted by $|z|$, for which he *postulated* what he wanted to get, and had to get to make any progress, namely,

$$|0'| = 0, |x| > 0 \text{ if } x \neq 0',$$
$$|zw| = |z||w|, \text{ and } |z + w| \leq |z| + |w|,$$

for all z, w in F.

This creative technique, frequently resorted to in modern abstraction, is reminiscent of Genesis: "God said, 'Let there be light,' and there *was* light"; Kürschák said, "Let there be absolute values," and there *were* absolute values. The outcomes were similar. The first fiat illuminated the world, the second lit up a good deal of mathematics with a new and unearthly light. To some alarmed conservatives it seemed that the end of the mathematical world had come. Others thought they saw a resplendent dawn breaking on a new universe. It may turn out that both were mistaken.

Notice that multiplication of absolute values is defined in the abstraction or generalization. A further generalization is achieved by dropping multiplication and attending only to the triangular inequality [8.9]. (R.) M. Fréchet (1878–) was the first (1906) to do this; he was shortly followed by many. He proposed $|x - y|$ as a definition of the *distance* between x, y. With the triangular inequality and some further reasonable requirements which need not be stated here, this definition preserves the familiar properties of the distance between two points in a plane. And so it goes.

Having mentioned distance I may as well state what some abstract geometers understand by the distance $D(p, q)$ between any two identical or distinct *elements* p, q, r, . . . of any class K. Abstracting the intuitive properties of distance as in a plane, the geometers lay down the following five postulates.

(1) To any elements p,q of K there corresponds a unique real number, $D(p, q)$, their *distance*.

(2) $D(p, p) = 0$.

(3) $D(p, q) \neq 0$ if p,q are distinct.

(4) $D(p, q) = D(q, p)$.

(5) $D(p, q) + D(q, r)$ is greater than or equal to $D(p, r)$.

The last is the abstraction of the triangular inequality; (4) says that the distance from p to q is equal to the distance from q to p; (2) makes the entirely reasonable assertion that the distance of a thing from itself is zero. And so on. Easy, is it not? Perhaps deceptively so.

A further step along the primrose path to abstraction was taken in 1922 by S. Banach (1892–1941), who had the honor and the misfortune to be Rector of the University of Warsaw when the Germans arrived in 1941 during World War II. (He died a natural death.) Banach started modestly enough from a class consisting of at least two completely arbitrary elements, and postulated that the class is closed under addition (as in a ring [5.3]) and under multiplication by real numbers [4.2] subject to certain simple, almost trivial, postulates. Again it seemed superficially to be the device of putting in what you want to get out. He also postulated an absolute value for which the triangular inequality holds. It might seem that from such slavish copying of the rudiments of elementary algebra and elementary geometry nothing really new could issue. But this appears to have been contradicted by experience. In spite of Banach's professional rivals who say that he put into his space the trivialities he wished to get out of it, 'Banach space' has at least unified the underlying concepts of several 'spaces' of interest to abstract geometers.

The examples of geometric abstraction described may appear too easy. They are, perhaps, after they have been done. The real difficulty is to decide in the first place what aspects of a theory can be profitably abstracted. It is indeed easy to abstract almost anything in technical mathematics, but without insight or luck the result is a barren set of postulates and a sterile list of formulas. The abstract algebraists, the abstract analysts, and the abstract geometers of the twentieth century have at least been lucky.

It seems to be the accepted etiquette of the abstractionists' game that the abstracter shall cover up his tracks. I know one prominent abstracter who spends many laborious months in working out detailed examples in an extremely difficult subject, and who then discards the examples, removes all traces of his exploratory investigations, and produces a beautiful and profound abstraction of the original theory that looks like a direct message from heaven. He then casually deduces in a few lines the theory from which he started—and but little else. This tactic, which is by no means infrequent, seems rather like carrying art for art's sake to excess.

However, abstraction was one of the outstanding landmarks of mathematics as it developed in the first half of the twentieth century. You may like it or you may not, but you cannot deny that it attracted a host of industrious and embarrassingly prolific workers. On the other hand you may agree with Hardy, England's leading mathematician of the half century before 1950, that what mathematics needs to keep it alive and healthy is a second Euler gifted with inventive imagination and not too greatly trammeled by meticulous rigor. But this again may express no more than a futile wish to sweep back the oncoming tide with a besom. We may take our choice.

Chapter 9

GROUPS

❖ 9.1 ❖ Multiplication Tables

In describing Klein's program for geometry we saw [8.1]
that the concept of a *group of operations* dominated at least
one major province of mathematics for half a prolific century. Groups also were found to be the structure behind
much of modern algebra, in particular the theory of algebraic equations. Wherever groups disclosed themselves, or
could be introduced, simplicity crystallized out of comparative chaos. Finally some modern philosophers became interested in this powerful, unifying mathematical concept
of groups as an important phase of scientific thought. As
the idea of a group was one of the outstanding additions to
the apparatus of scientific thought since Galois coined the
term in 1831, I shall discuss it at some length. Incidentally
I shall pick up and amplify the details from groups cited in
earlier chapters. These will be signalized as we pass them.

Before proceeding to the official definition of an abstract
group, I add a word of caution. Vast as was the panorama
swept in from the vantage points of groups, it was by no
means relevant for the whole of mathematics, either ancient
or modern, although as late as the 1900s some enthusiasts asserted that all mathematics worth cultivating was a succession of episodes in the theory of groups. In many a fertile
mathematical province groups either play no part or play
only a very subordinate one. When *Principia mathematica*
began coming out in 1910, Russell almost exulted that
groups were of practically no significance in mathematical

logic. The whole theory of groups itself is but an incident, though an impressive one, in the algebra of the nineteenth and twentieth centuries.

Groups are first subdivided into two grand divisions, *finite* and *infinite*. The number of distinct operations in a finite group is finite; in an infinite group the number of distinct operations is infinite. The subject was extensively developed in the nineteenth century by a host of mathematicians, among whom Galois, A. L. Cauchy (1789–1857), Jordan, M. S. Lie (1842–1899), and L. Sylow (1832–1918), may be mentioned.

A *finite* group according to a famous dictum of Cayley's in 1854 is defined by its *multiplication table*. Such a table states completely the laws according to which the operations of the group are combined. Here is a specimen which can be easily understood.

$\times$	I	A	B	C	D	E
I	I	A	B	C	D	E
A	A	B	I	D	E	C
B	B	I	A	E	C	D
C	C	E	D	I	B	A
D	D	C	E	A	I	B
E	E	D	C	B	A	I

This group contains the six operations, I, A, B, C, D, E. I shall state what the table says about any pair of these operations, say B and D. Take any letter, say B, from the left-hand vertical column, and any letter, say D, from the top horizontal row, and see the entry C in the table where the B row and the D column intersect. It is just as if we were to *multiply* B by D, say $B \times D$, and get the answer C. Instead of writing $B \times D$, we shall write BD, which says to take B from the *left*, D from the *top*, and find where the corresponding row and column intersect. This gives the result C; so we write $BD = C$.

What about DB, found according to the same rule? It is not equal to C, but to E; namely, $DB = E$. So in this kind of *composition*, BD and DB are not necessarily equal. The reader may easily satisfy himself that although the commutative law has gone, the associative is still valid. For instance $(AB)C = A(BC)$.

Let now x be *any* member of a given class on which I, A, B, C, D, E operate. We *postulate* that the result of operating with any one of I, A, B, C, D, E *on x gives another member of the class*. Let us write $B(x)$ (read, 'B on x') for the *result* of operating on x with B. By our postulate this is some member of the given class, so we can operate on $B(x)$ with D. The result is written $BD(x)$, which is again in the class. Now, the assertion of the table that $BD = C$ says that, instead of performing the operations B, D *successively*, first B and then D, we could reach the same final result in *one* step, by performing the operation C on x. Thus, *the class is closed with respect to the operations I, A, B, C, D, E.* For the results of performing the operations of the set successively are always in the set. If the reader doubts this, let him follow the rule which gives $BD = C$, $DB = E$, $CE = A$, $EC = B$, and so on, and try to escape from the table. Lay aside this book for a moment and reflect on the miracle that such closed, finite sets actually exist.

Notice the effect of operating with I. The table says that $AI = IA = A$; $BI = IB = B$, and so for all. Thus I as an operation changes nothing; it is called the *identity*. This is the operation that was occasionally overlooked [8.1] before the postulates for a group were explicitly stated and universally accepted.

The last thing to be observed attentively is this. Given *any one* of I, A, B, C, D, E, say X, there is *always exactly one other of* the six, say Y, such that $XY = I$. Further, for every such pair X,Y, it is true that $XY = YX$. It is not asserted that X,Y are *necessarily* distinct. For example, if X is the particular operation B, then the table says that $Y = A$, because $BA = I$; if X is E, then Y also is E. Two operations X,Y such that XY is the identity are called *inverses* of one another. The table states that *each element of the set has a unique inverse.* This also was sometimes ignored or tacitly assumed, as we observed [8.1] in discussing Klein's program for geometry.

A set of operations having all the foregoing properties is called a *group*. The definition by postulates will be given presently.

For the moment let us see that an instance of the group defined by the specimen multiplication table actually exists. There are dozens of examples—all in different parts of mathematics. Here is a very simple specimen from arithmetic. Start with any number different from o, 1, say x. We can *subtract x from* 1, and we can *divide* 1 *by x*, getting the new numbers $1 - x$ and $1/x$. Repeat these operations on the new numbers. Then $1 - x$ gives back x and a new number $1/(1 - x)$; $1/x$ gives the new number $1 - 1/x$ or $(x - 1)/x$, and gives back x. If you keep this up forever, you can get but one or another of the six numbers x, $1/x$, $1 - x$, $1/(1 - x)$, $(x - 1)/x$, $x/(x - 1)$. Now let I be the operation which transforms x into itself, $I(x) = x$; let B be the operation which transforms x into $(x - 1)/x$, or $B(x) = (x - 1)/x$; and so on, with $C(x) = 1/x$,

$$D(x) = 1 - x,$$

$E(x) = x/(x - 1)$. A little patience will show that *these I, A, B, C, D, E* satisfy the multiplication table. Another instance will be given shortly [9.4] when we come to substitution groups.

The number of different operations in a group is called its *order*. Thus our group is of order 6. Looking at the table more closely, we see several smaller groups, called *subgroups*, within the whole group; for example, those whose multiplication tables are

×	I
I	I

×	I	C
I	I	C
C	C	I

×	I	A	B
I	I	A	B
A	A	B	I
B	B	I	A

whose respective orders are 1, 2, 3. Now 1, 2, 3 are *divisors* of 6, and we have illustrated a fundamental theorem of groups, the *order of any subgroup of a given finite group is a divisor of the order of the group.*

The following postulates for a group should now be intelligible.

We consider a class and a rule, written as ∘, by which the two things A,B in any ordered couple (A, B) of things in the class can be combined so as to yield a unique thing which is again in the class.

The result of combining A,B in the ordered couple (A, B) where A and B are any things in the class, is written $A \circ B$.

POSTULATE (*Closure under* ∘). If A,B are in the class, then $A \circ B$ is in the class.

POSTULATE (*Associativity of* ∘). If A,B,C are in the class, then $(A \circ B) \circ C = A \circ (B \circ C)$.

POSTULATE (*Inclusion of identity*). There is a unique thing I in the class such that $A \circ I = I \circ A = A$ for every thing A in the class.

POSTULATE (*Unique inverse*). If A is any thing in the class, there is a unique thing, say A', in the class such that $A \circ A' = I$.

The foregoing postulates define a *group:* the *class* is said

to be *a group under* (or with respect to) *the composition* ○.
The postulates contain redundancies but are more easily
seen in the above inelegant form. The $A,B,C, \ldots$ are our
previous 'operations.' For simplicity we shall write $A \circ B$
as AB.

It is instructive to compare the postulates for a group
with those for a field [3.1]. It will be seen that, if we *suppress
the commutative property* of multiplication in a field, the
remaining postulates for multiplication are those of a
group, and likewise for addition.

If the composition ○ does have the commutative property
(as in the arithmetical examples above), the group is called
commutative, or *abelian* (after Abel).

Before leaving the table, let us note a few simple exercises. *Powers* of an operation are defined as successive per-
formances of the operation; thus, A^2 means AA, A^3 means
AAA, and so on. Since there are only a finite number (six)
of operations in the specimen table, and the table is closed,
the successive powers of each operation must sooner or
later repeat, from which it follows easily that for each operation different from I, say for A, there is a least positive
integer, say a, greater than 1, such that $A^a = I$, and
$A, A^2, \ldots, A^a$ are all distinct; a is called the *order* of A.
From the table, each of A,B is of order 3, and each of
C,D,E is of order 2,

$$A^3 = B^3 = I, \ C^2 = D^2 = E^2 = I,$$

and, for example, I,A,A^2 are all distinct and form a subgroup of the original group. This subgroup, being generated
by the powers of A is called *cyclic*—all its operations form
a single cycle. Generally, if the element X of a finite group
is of order x, the powers of X generate a cyclic group of
order x,

$$I, X, X^2, \ldots, X^{x-1}.$$

A cyclic group is necessarily commutative.

The entire table need not be written out to give the whole group. For our example it can be verified that the following relations generate the group:

$$A^3 = I, \; C^2 = I, \; B = A^2, \; D = AC = CA^2, \; E = A^2C = CA.$$

❖ 9.2 ❖ *Isomorphism, Homomorphism*

Isomorphism and homomorphism were defined for rings [5.4], and it was remarked that precisely similar definitions hold for systems closed under a single binary operation. A group is such a system, so it may be unnecessary to repeat for groups what was said in connection with rings. But as isomorphism and homomorphism, or multiple isomorphism, are of such great importance for groups, I shall give independent definitions of the relevant concepts. Historically, these concepts originated in the theory of finite groups.

Let G,G' be two groups of the same order n. If there is a one-one correspondence between the operations or *elements* of G and those of G' such that to every operation of G there corresponds a unique operation of G', and to every operation of G' there corresponds a unique operation of G, and moreover to the product AB of any two operations A,B of G there corresponds the product $A'B'$ of the operations corresponding to A,B respectively, G,G' are said to be *simply isomorphic*.

It is clear that simply isomorphic groups are only trivially distinct. A given group may have innumerable specific interpretations. Behind them all is a single *abstract group* which may be defined by its multiplication table. In enumerating finite groups, simply isomorphic groups are not counted as distinct. To exhibit an example I shall have to draw on [9.4], Substitution Groups, and ask the reader who may be interested to look back after he has read that section. Let i denote the imaginary unit $\sqrt{-1}$ of complex numbers [4.2] with its usual properties, $i^2 = -1$, $i^3 = -i$, $i^4 = 1$. Take for G the group $I(= 1)$, i, -1, $-i$, and for G' the group I' (the identity), $(abcd)$, (ac), (bd), $(adbc)$ with the one-one correspondence [5.4]

$$I \leftrightarrow I',$$
$$i \leftrightarrow (abcd),$$
$$-1 \leftrightarrow (ac)(bd),$$
$$-i \leftrightarrow (adbc).$$

By constructing the multiplication tables (or otherwise) it is seen that G,G' are simply isomorphic.

Homomorphism, or *multiple isomorphism*, for groups is defined as follows.

Let G,G' be two groups, the order of G' being *less than that of* G. If to each operation S of G there corresponds a single operation S' of G', while to the product ST of the operations S,T of G there corresponds the operation $S'T'$ of G', the group G is said to be *multiply isomorphic with* (or *to*) G.

The most important instances of multiple isomorphism occur in connection with *normal* (or *self-conjugate*, or *invariant*) *subgroups* of a given group. A normal subgroup separates the entire group into congruence classes in a sense abstractly identical with that already described [5.4] for rings.

❖ 9.3 ❖ Complexes, Cosets, Normal Subgroups

As these concepts offer several illustrations of structure as it occurs in groups, I shall describe them in some detail. The elements of the group G will be denoted by small letters a, b, c, . . . , g, h, . . . , the inverses of these by a^{-1}, . . . , g^{-1}, h^{-1}, . . . ; complexes and, presently, subgroups, by script capitals, $\mathcal{G}$, $\mathcal{H}$, ... A *complex* $\mathcal{G}$ is any set of elements of G, and the *product* $\mathcal{G}\mathcal{H}$ of two complexes is defined as the set of all products gh where g is in $\mathcal{G}$ and h is in $\mathcal{H}$. If $\mathcal{G}$ is the single element g, we write $g\mathcal{H}$, and similarly for $\mathcal{G}h$, where $\mathcal{H}$ is now the single element h. Under this definition of multiplication for complexes the associative law holds, $\mathcal{G}(\mathcal{H}\mathcal{K}) = (\mathcal{G}\mathcal{H})\mathcal{K}$, so that either product may be written unambiguously as $\mathcal{G}\mathcal{H}\mathcal{K}$. Since G is closed under the 'multiplication' of the group, $\mathcal{G}\mathcal{G} = \mathcal{G}$. If $\mathcal{G},\mathcal{H}$ are *subgroups* of G, it is readily seen that $\mathcal{G}\mathcal{H}$ is also a subgroup of G if, and only if, $\mathcal{G},\mathcal{H}$ are permutable, that is, if and only if $\mathcal{G}\mathcal{H} = \mathcal{H}\mathcal{G}$. If G is an abelian (commutative) group, this condition is automatically satisfied.

What follows goes back to Galois (1831), although the phraseology differs from his. If $\mathcal{G}$ *is a subgroup* of G and a is any element of G, $a\,\mathcal{G}$ is called a *left coset* of G, and $\mathcal{G}\,a$ a *right coset* of G. If a is in $\mathcal{G}$, $a\mathcal{G} = \mathcal{G}a = \mathcal{G}$. An alternative terminology for cosets is more suggestive, *residue classes* (compare [5.4]). By a few simple changes in the wording anything said for left cosets holds also for right cosets, so it will suffice to state results only for the former. The proofs of the following statements are almost immediate. If $a\mathcal{G} = b\mathcal{G}$ it is sufficient but not necessary that $a = b$, for if $a^{-1}b$ is in $\mathcal{G}$ the equality holds. Cosets having no element in common are said to be *distinct*. The element a of G is in the coset $a\mathcal{G}$, and a is called the representative of $a\mathcal{G}$—in analogy with the like for residue classes [5.3]. A coset formed from a given subgroup $\mathcal{G}$ is a group only when the coset is G itself, that is, when the coset is eG, where e is the identity element of G. The *index in* (or *under*) G of the subgroup $\mathcal{G}$ is the number of distinct cosets of $\mathcal{G}$; it is equal to the quotient of the order of G by the order of $\mathcal{G}$. From this it follows that the order of any subgroup of G is a divisor of the order of G. Examples of this theorem were given earlier [9.1].

We come now to *normal subgroups,* or *invariant subgroups,* or *normal divisors,* of a group G. This concept was introduced by Galois in his theory of algebraic equations, where it is of fundamental importance. We shall recur to this. If for every element a in G the subgroup $\mathfrak{g}$ is such that its left and right cosets are equal, $a\mathfrak{g} = \mathfrak{g}a$, $\mathfrak{g}$ is called a *normal divisor* of G. It follows at once that the product $a\mathfrak{g}b\mathfrak{g}$ of two normal divisors is the coset $ab\mathfrak{g}$.

We now connect these ideas with the general concept of automorphism as previously described [9.2]. Briefly, an *inner* automorphism of a group G is a mapping of the elements of G onto the elements of G by means of a correspondence of the following kind. If x is any element of G, and a is a fixed element of G, the correspondence is $x \rightarrow x'$, where $x' = axa^{-1}$; x' is called the *transform of x by a*; x' and x are called *conjugate* elements. The reason for the name is that if x' is conjugate to x, then x is conjugate to x'. For, from $x' = axa^{-1}$, by multiplying throughout on the left by a^{-1} and on the right by a, we get $a^{-1}x'a = x$, or $x = a^{-1}x'a$, which is of *the same form* as $x' = axa^{-1}$ with a replaced by a^{-1}, since $(a^{-1})^{-1} = a$. The inner automorphism $x \rightarrow axa^{-1}$ applied to the elements of a subgroup $\mathfrak{g}$ of G takes $\mathfrak{g}$ into the subgroup $a\mathfrak{g}a^{-1}$, called a *conjugate subgroup*. In the special and important case of $\mathfrak{g}$ a normal divisor, all the conjugates of $\mathfrak{g}$ are identical with $\mathfrak{g}$, that is, $\mathfrak{g}$ is transformed into itself, or is left invariant, by all the inner automorphisms.

A group may have automorphisms other than the inner; these are called *outer* automorphisms. They need not concern us here.

To bring out the analogy between residue classes for rings and their ideals [5.6] and the separation of a group into cosets with respect to an invariant subgroup (normal divisor), I return to multiple isomorphism, or homomorphism [9.2].

It should be noted that a group may have no invariant subgroups besides itself and the identity element. Such a group is called *simple*.

If the groups G, G' are multiply isomorphic as defined [9.2], G' is called a *homomorphic mapping of G on G'*. All the elements of G that map into a fixed element of G' constitute a class under the homomorphism, and from the definitions it is immediate that an element of G can belong to only one class. All the elements of G are thus separated into mutually exclusive classes by the homomorphism, and the identity element of G maps into the identity element of G'; also, elements of G that are inverses of each other map into elements of G' that are inverses of each other. It can be shown quite simply that the class of all·those elements of G that map into the identity element of G' is a normal divisor of G, and that the remaining classes are the cosets of this normal divisor.

Suppose now that we proceed from a normal divisor $\mathfrak{g}$ of G and the cosets $a\mathfrak{g}$, $b\mathfrak{g}$, ... We saw that $a\mathfrak{g}b\mathfrak{g} = ab\mathfrak{g}$. From what precedes it follows that the cosets form a group, called the *factor group*, or *quotient group* of G with respect to $\mathfrak{g}$, written $G/\mathfrak{g}$. An example of a quotient group is given in [9.5]. As the conclusion of this whole matter, I shall state the *homomorphism theorem* for groups. Its proof is not too far to seek if the foregoing definitions and theorems are kept in mind. Let $\mathfrak{g}$ be that normal divisor of G which in the homomorphic mapping of G into G' maps into the unity element of G'. Then G' is isomorphic to $G/\mathfrak{g}$. There is a converse which I omit.

The rather large number of technical terms whose meanings must be remembered, though perhaps less now than it was in the nineteenth-century work, makes the theory of groups look harder than it is. The details described here are only a small sample. They have been chosen partly for their own interest, partly to make the very brief account of the Galois theory for abstract fields [9.8] intelligible when we come to it.

❖ 9.4 ❖ *Substitution Groups*

How many distinct finite groups of any given order are there? The Lord knows. This question absorbed the working lives of dozens of algebraists during the late nineteenth century and well into the twentieth. A moment's thought will show that for any given integer n the number of multiplication tables defining a group must be finite. But even for a table constructed on only twenty letters the labor is considerable. The question had been practically dropped by 1915, possibly because some algebraists began to suspect that it should never have been asked in the first place. Nevertheless, nearly all the finite groups of scientific (as opposed to mathematical) interest had been worked out long before the modern quantum theory initiated in 1925 found a use for them.

An aid in this work was the theory of *substitution*, or *permutation*, groups, which give a concrete representation of any finite group such as the specimen exhibited in [9.1]. This mode of representation provided tangible material in place of abstract postulates for algebra to take hold of. The connecting link was W. von Dyck's (1856–1934) theorem (1882): *Every group of finite order n can be represented as a group of substitutions on n letters.* I shall not reproduce the proof. We must be content here to see what substitutions in the present technical sense are, and then indicate how they and their groups became of unforeseen physical significance.

The meaning of the terms will be clear from a simple

example. There are precisely 6 possible orders in which the 3 letters *a,b,c* can be arranged, namely,

$$abc, \ acb, \ bca, \ bac, \ cab, \ cba.$$

Our object is to define a set of operations, called *substitutions*, which will enable us to pass from any one of these arrangements to any other.

To pass from *abc* to *acb* we leave *a* as it is, replace *b* by *c* and *c* by *b*. This operation is written (*bc*), which may be read '*b* into *c*, and *c* into *b*.' In the same way (*abc*) is read '*a* into *b*, *b* into *c*, and *c* into *a*.' Applied to the *arrangement* *abc* the *substitution* (*abc*) yields the new *arrangement* *bca;* applied to *cab* it gives *abc*.

Take any one of the 6 arrangements of *a*, *b*, *c* as the initial one, say *abc*, and under each of the others write the substitution which changes *abc* into that one:

abc,	*acb,*	*bca,*	*bac,*	*cab,*	*cba,*
I	(*bc*),	(*abc*),	(*ab*),	(*acb*),	(*ac*).

The letter *I* stands for the *identity substitution*, and may be interpreted as '*a* into *a*, *b* into *b*, *c* into *c*,' namely as the substitution which does not change *any* arrangement of *a,b,c*.

A moment's reflection will show that the 6 substitutions above must form a group, since the effect of any one of them upon the arrangement *abc* is either to leave *abc* unchanged (when operated on by *I*) or to change *abc* into one of the other 5 possible arrangements. So if one substitution be followed by another, operating on a given arrangement, the total effect is equivalent to that of some single substitution in the set, since all possible arrangements are included in the set. Similarly to each substitution corresponds a unique substitution, the same as or different from the given one, which will restore the initial order *abc* after the given substitution has operated. Thus every substitution in the set has a unique inverse.

Let S, T be any substitutions. The effect of S *followed* by T is written ST. That is, the substitution which operates *first* is written *first* (here, but some authors use the other way; it makes no material difference which convention is used, provided we remember). If the effect of ST (S *followed* by T) is equivalent to the effect of U, we write $ST = U$. As a matter of convenience, we call ST the *product of S and T in this order.*

We shall now see that ST and TS are not necessarily the same. Indeed common sense would suggest that they are likely sometimes to be different. For ST is 'S followed by T,' while TS is 'T followed by S,' and this world would be much crazier than it already is if it made no difference in what order operations were performed on an operand. You cannot, for instance, digest your dinner before you have eaten it. To see the point for substitutions we have

$$(abc)(ac) = (ab), \ (ac)(abc) = (cb).$$

Enough has now been said about the 'multiplication' of substitutions to make the verification of the following easy. Denote each of the substitutions other than I by the single written under it,

(abc)	(acb)	(bc)	(ac)	(ab)
A	B	C	D	E

Then I, A, B, C, D, E form the group whose multiplication table is given in [9.1]. Such finite groups of substitutions are called *substitution groups,* or *permutation groups.*

Two particular groups, each of substitutions on any number n of letters, have received special names on account of their importance in applications, particularly in the theory of algebraic equations. There are precisely $n!(= 1 \times 2 \times 3 \times \cdots \times n)$ possible different arrangements of n letters. The substitution group corresponding to these is therefore of order $n!$. It is called the *symmetric* group on n letters, because each of its substitutions leaves invariant any symmetric function of the n letters. Every substitution group on n letters is a subgroup of the symmetric group, but this does not get us very far in making a census of groups. We shall describe the other special group shortly [9.5].

❖ 9.5 ❖ *Interpretations*

I have defined substitution groups and have just exhibited one such as an instance of the table in [9.1]. The table itself defines an *abstract group*, so called because, beyond the law according to which the symbols I, A, B, . . . are combined, in accordance with the postulates for a group, the symbols are entirely arbitrary. When, as in the instance of substitutions, a specific interpretation consistent with the postulates is assigned to the symbols, we have a *representation* of the group. Each representation of a group gives an interpretation or application of it. There are several ways of representing finite groups, but that by substitutions is sufficient here.

One of the earlier applications of groups was to the study of crystal structure. With the X-ray analysis of crystals, this application was revived and greatly elaborated. It is purely a matter of geometrical classification, and although extremely useful in cataloguing the various possible types of crystals, it cannot be said to have any very profound physical significance. Of a totally different order is the next.

That anything so simple and so abstract as finite groups should be useful in the mathematical exploration of atomic structure and spectra seems like a miracle. Yet it is true.

The connection between nature and groups is made through an apparent triviality. Ignoring the subtle metaphysical implications of their assumption, physicists assume that two electrons are physically indistinguishable from one another. Hence if in a configuration of electrons any two of them be interchanged the configuration will be the same physically as it was before the interchange. That is, in the language of Chapter 6, the configuration in *invariant* under all interchanges of the members of a pair of electrons. From this it is easily seen that the configuration is invariant under *all* permutations of its electrons, since *any* permutation is a product of transpositions as noted immediately.

Suppose for definiteness that precisely 3 electrons, e_1, e_2, e_3, occur in the configuration. As in [9.4], denote by $(e_1 e_2 e_3)$ the substitution which replaces e_1 by e_2, e_2 by e_3, and e_3 by e_1, and similarly for $(e_1 e_2)$, the substitution which replaces e_1 by e_2 and e_2 by e_1—and so on for the rest. Now $(e_1 e_2)$ interchanges e_1 and e_2. Hence, by our physical assumption, $(e_1 e_2)$ is an operation which leaves the configuration invariant. A substitution, such as $(e_1 e_2)$ which contains precisely two letters, is called a *transposition*. It is easily proved that any substitution containing more than two letters can be written as a product of transpositions. Thus all the substitutions in the group

$$I, (e_1 e_2 e_3), (e_1 e_3 e_2), (e_2 e_3), (e_1 e_3), (e_1 e_2),$$

which is the same group (abstractly) as that considered in [9.1], as may be seen when e_1, e_2, e_3 are replaced by a, b, c, leave the configuration invariant.

The rule for decomposing a substitution into a product of transpositions is clear from the examples

$$(abc) = (ab)(ac), \ (abcd) = (ab)(ac)(ad),$$
$$(abcde) = (ab)(ac)(ad)(ae),$$

which can be verified by multiplying out the right-hand members.

A substitution which is decomposable into an even number of transpositions is called *even*, for example (abc) and $(abcde)$. Exactly one-half of all the substitutions of the symmetric group [9.4] on n letters are even. They form a normal subgroup of the symmetric group. This is the other special group mentioned in [9.4]. It is of order $n!/2$, and is called the *alternating group* on n letters.

The promised [9.2] example of a quotient group may be given here, as it refers to the alternating group G on 4 letters a,b,c,d which has the normal subgroup H of order 4,

$$H = I, \; (ab)(cd), \; (ac)(bd), \; (ad)(bc),$$

as the reader may easily verify. The cosets are

$$(abc)H = (abc), \; (bdc), \; (adb), \; (acd),$$
$$(acb)H = (acb), \; (adc), \; (bcd), \; (abd),$$

so that G may be written as in the first column of the mapping

$$H \to I',$$
$$(abc)H \to (xyz),$$
$$(acb)H \to (xzy),$$

of G on the cyclic group I', (xyz), (xzy) on three letters x,y,z, which is the quotient group G/H. The verification of this will provide exercises in several of the concepts described earlier.

Although the model of the atom as a central nucleus surrounded by planetary electrons—like a miniature solar system—has long ceased to be adequate for atomic physics and spectroscopy, it is sufficient here to bring out the point: with the atom is associated a substitution group under which the atom is (physically) invariant. A little more precisely, the possible configurations of the electrons about the nucleus are invariant under all permutations of the electrons. For n electrons the symmetric group on n letters is suggested.

From this apparent triviality it is possible, by applying the full machinery of the theory of groups, to give a coherent picture of much of quantum mechanics and the theory of spectra. I shall not attempt here to describe what the modern quantum theory is about. It is sufficient to say that since 1925 it has revolutionized our conceptions of matter, atomic physics, spectroscopy, and the metaphysics of physical science. Several accounts for the general reader are readily available in as untechnical language as is feasible. Here I have simply pointed out another instance of the services which mathematics, nourished by mathematicians for its own sake, has rendered the sciences.

To the regret of algebraists this beautiful application of their work quickly passed out of fashion when it was discovered (1926) that the mathematics of the quantum theory can be done without groups in a manner more familiar to working physicists. But the group (and matrix [6.3]) method was the first, and may have suggested the technique of boundary-value problems more congenial to physicists.

It may be mentioned in passing that the theory of groups was applied by Eddington to atomic structure in his relativistic theory of protons and electrons (1936), and again in his posthumously published (1946) *Fundamental theory* of the physical constants. Physicists took kindly to neither. His famous '137' for the fine-structure constant of spectroscopy came out of groups by way of physics and metaphysics. He got 136 (incorrect) first because he overlooked the fact that a group has an identity element.

I may refer here to Sylvester's prophecy [6.6].

❖ *9.6* ❖ *Infinite Groups*

A word must be said about infinite groups. These again fall into two grand divisions. In the first, the distinct things are denumerable [4.2]; that is, the things in the group can be counted off 1, 2, 3, . . . , but *we never come to the end.* Such groups are infinite and *discrete.* In *continuous* groups the number of distinct things is infinite, but not denumerable; the things can *not* be counted off 1, 2, 3, . . . , but are as numerous as the points on a line [4.2].

Continuous groups arise in the following way, among others. In school geometry it is *assumed,* as we noted [8.5], that a plane figure, say a triangle, can be moved about all over the plane and retain its shape (size of angles and length of sides). Consider the group of all motions of a rigid figure in a plane. Evidently the group contains *infinitesimal* transformations, for we can shift the figure from one position to another by stages as small as we please.

Another example of a group consisting of infinitesimal transformations is that of the rotations of a rigid, solid body about a fixed axis. Either the body as a whole may be thought of as being moved from one position to another, or the motion may be realized by subjecting each individual point to an appropriate transformation. Both points of view are useful. We shall have considerably more to say about this in a later chapter.

Now let us recall that the equations of mechanics and those of classical mathematical physics are *differential equations.* Examples will be given in Chapters 14, 15, and 16. Roughly, such equations express laws concerning rates of change of one or more continuously varying magnitudes with respect to one or more others. As a simple example, the velocity of a falling body is the 'rate' of change of position with respect to time. The vast theory of differential equations was greatly furthered by the introduction of continuous groups into its study. For instance, the central equations of higher dynamics—those named after their discoverer, Hamilton—when viewed from the standpoint of continuous groups become much clearer than before. All this long since became part of a standard course in analytical mechanics for advanced students.

The study of such groups absorbed the working lives of many mathematicians from 1873 to the early years of the twentieth century, when interest diminished, owing to a great memoir published in 1894 by E. Cartan (1869–), which disposed of several of the main problems. After a long neglect following the death in 1899 of Lie, the creator of the theory of continuous groups, certain aspects (particularly the algebraic) of the theory again attracted creative mathematicians. 'Lie algebras,' as they are called, were added to the stock, already large, of special algebras. To make their introduction seem reasonable I should have to explain certain details of Lie's theory, which would be too much of a digression here. The great unifying power of continuous groups had long been familiar in the classical algebra of invariants, linear associative algebra, and theoretical mechanics. With the new physics, beginning with general relativity in 1915 and continuing with the quantum mechanics of 1925, continuous groups suddenly were seen to be of fundamental importance in the description of nature. New and yet more general geometries, suggested partly by physics, were created in swarms, and in this outbrust the theory of continuous groups was at least a highly suggestive guide. In this renaissance of infinitesimal geometry, Cartan was a leader.

❖ 9.7 ❖ The Icosahedron

Although it is not my intention to discuss special results, I may close this description of groups by referring to one which would have delighted Pythagoras and have caused him to sacrifice at least a thousand oxen to his immortal gods. The story covers nearly 2,200 years. Only the high points can be indicated.

The Greek geometers early discovered the five regular solids of Euclidean space—the tetrahedron, cube, octohedron, dodecahedron, and icosahedron, of 4, 6, 8, 12, and 20 sides respectively—and proved that there are no others. This discovery begot much of the incredible mysticism of later and less exact thinkers.

Our next high point is about 2,000 years farther on. For over two centuries algebraists had tried in vain to solve the general equation of the fifth degree,

$$ax^5 + bx^4 + cx^3 + dx^2 + ex + f = 0,$$

until Abel in 1826 and Galois in 1831 *proved* that it is impossible to express x by any combination of the given numbers a,b,c,d,e,f, using only a *finite* number of additions, multiplications, subtractions, divisions, and extractions of roots. Thus it is *impossible to solve the general equation of the fifth degree algebraically.* On the eve of that stupid duel in which he was killed, Galois, then in his twenty-first year, wrote out his mathematical testament, in which, among other tremendous things, he sketched a great theorem concerning all algebraic equations. He reduced the problem of the *algebraic* solution of equations to an equivalent, approachable one in groups. As this is an outstanding landmark in algebra, I shall state Galois's theorem, in the hope that some may be induced to go farther and find out for themselves exactly what it means: *an algebraic equation is algebraically solvable, if, and only if, its group is solvable.* No more technical knowledge is necessary to follow the proof than is possessed by high-school graduates. But the proof is not easy. As a consequence of this perfect theorem, it is *impossible to solve the general equation of any degree greater than the fourth algebraically.*

In 1858 Hermite *solved the general equation of the fifth degree,* not algebraically, for that would have been to do the impossible, which is too much even for mathematicians, but by expressing x in terms of elliptic modular functions (a sort of higher species of the familiar trigonometric functions).

Our last peak was discovered by Klein, who showed in 1884 that the profound work of Hermite was all implicit in the properties of the group of rotations about axes of symmetry which change an icosahedron into itself—that is, which twirl the solid about so that, say, a given vertex slips over to the place where some other vertex was, and so for all in every rotation. There are sixty such rotations.

That the rotations of an icosahedron and the general equation of the fifth degree should be unified from the higher standpoint of groups, is a good illustration of the power of the concept of an abstract group.

The far-reaching power of the theory of groups resides in its revelation of identity behind apparent dissimilarity. Two theories built on the same group are structurally identical. The more familiar is worked out; the results are then interpreted in terms of the less familiar.

❖ 9.8 ❖ Galois Theory

A very abstract little book, *Galois Theory,* was being read by a class in algebra, all of whose members were acquainted with the classical form of the Galois theory as it was usually presented before the advent of modern abstract algebra in the 1920s. When the book had been digested, one student wondered what Galois would have thought of it all. The consensus was that he would have had difficulty in recognizing it as anything to which his name might be legitimately attached. Another student said Galois would at least recognize his own name on the cover. "Oh no, he wouldn't," another objected. "He would say, 'Who the hell is this guy Theory?'"

To indicate some use for a few of the general ideas described earlier, I shall describe what the Galois theory for an abstract field is about. We must go back to Chapter 5 and pick up what was said about algebraic extensions [5.7] of a field F. The roots of an irreducible polynomial in F are distinct, or *separable*. Adjunction to F of the roots of an irreducible polynomial is said to give a *separable extension*. An algebraic extension E of F is called *normal*, or *Galoisian*, if every irreducible polynomial in F which has a root in E has all its roots in E. Such an extension is obtainable, for example, by adjoining to F all the roots of an irreducible polynomial. If E is an arbitrary normal extension of F, the automorphisms [5.4] of E which leave the elements of F fixed form a group, say G, called the *Galois group* of E with respect to F. To every subfield E' between F and E corresponds a unique subgroup of G, which I shall denote by $G(E)$, consisting of all those automorphisms of G leaving all the elements of E' fixed. Conversely to every subgroup G' of G there corresponds a unique subfield of E which I shall denote by $E(G')$, formed of all those elements of E which remain fixed under the automorphisms of G'. To unscramble the following formula, note that G outside a parenthesis refers to a group, E to a field. For instance $G(E(G'))$ is the subgroup of the group G leaving fixed the elements of the subfield $E(G')$ defined above. Similarly for $E(G(E'))$. If there is a one-one correspondence between subfields and subgroups such that

$$G(E(G')) = G', \ E(G(E')) = E',$$

E is said to have a Galois theory. The necessary and sufficient condition that this be so is that E be a finite separable extension of F.

In the older presentation this was the correspondence between subfields and subgroups leaving the elements of successive ground fields fixed. The connection with the solvability of algebraic equations by radicals is through the structure theory of groups, in particular the Jordan-Hölder (O. Hölder, 1859–1937) theorem (1869, 1889) on series of composition for groups. This has been recast and generalized (as what has not?) in many ways, including that by lattice theory. I shall content myself—and, I hope, the reader—with a single statement which I know to be true. It is from the G. Birkhoff's *Lattice theory* (2d ed., p. 88, 1948): "There is an enormous literature on this result."

Chapter 10

A METRICAL UNIVERSE

❖ 10.1 ❖ *From Pythagoras to Descartes*

'Metrical' in the heading has no connection with what
Pythagoras might have meant by it when he lost himself in
the mysterious harmonies of the music of the celestial
spheres. It refers only to *measurement* as that basic concept
occurs in geometry, everyday life, and relativity. This
should be enough for anyone.

I recall that Proposition 47 in Book I of Euclid's *Elements* of approximately 300 B.C. asserts that the square on
the longest side of any right-angled triangle is equal to the
sum of the squares on the other two sides. This famous
theorem, usually attributed to Pythagoras [1.2] in the sixth
century B.C., is the taproot of both the ancient and the
modern theories of *metrics*—measurement—as it appears
in geometry and other departments of mathematics; for
example, abstract analysis [8.9]. Years ago I had a pamphlet
with over forty different proofs of this, the 'Pythagorean
theorem,' all highly ingenious. The traditional figure usually given in school geometries recalls some garment that
might be hung up to dry on the Monday-morning clothes-
line, and Russian schoolboys, before the consecrated pedagogues of the New Order took over the indoctrination of
the young, used to call the figure "Pythagoras' pants."
Such levity is, I understand, no longer condoned.

The Pythagorean theorem is one of the enduring landmarks of all mathematical history, not only for its simplicity, generality, and intrinsic beauty, but also for what

it has inspired in current mathematics, from geometry to analysis. As the theorem is elementary and fundamental, some indication of a proof should be given before we pass on to its generalizations and their implications. My precious pamphlet having long since been borrowed, I reproduce the skeleton of a visual proof given (1945) by H. Baravalle. I thank the editors of *Scripta Mathematica* (1947) for permission to reproduce this here. The reader may convince himself, with a pair of scissors if necessary, that the figure really does make the theorem *intuitively* evident. From there on out, a *formal* proof is a matter of easy routine. The figure is shown on page 190.

Before bidding Pythagoras hail and farewell for the last time, we should note that his famous theorem probably was known, but without proof, to the Babylonians of the second millennium B.C., or about 1,300 years before Pythagoras was born. It may even have been observed by the Sumerian predecessors of the Babylonians. Research since 1940 in the ancient history of mathematics has shown that the Babylonians of about 1900 B.C. were familiar with universal rules for finding solutions in integers x,y,z of $x^2 + y^2 = z^2$, such as $x = 3$, $y = 4$, $z = 5$, and $x = 5$, $y = 12$, $z = 13$, and there is some evidence that they knew the general Pythagorean theorem in its geometrical form. The historical blank from the Babylonians to Pythagoras has yet to be filled. We can only speculate as to where all these promising beginnings of mathematics disappeared in the centuries from Babylon to early Greece.

A singular feature of Babylonian mathematics is the total anonymity of the men responsible for any of its great discoveries. To a modern mathematician, aware of the acute professional jealousies and greed for personal glory among some of his colleagues, this disregard for scientific notoriety with its attendant emoluments in academic prestige and

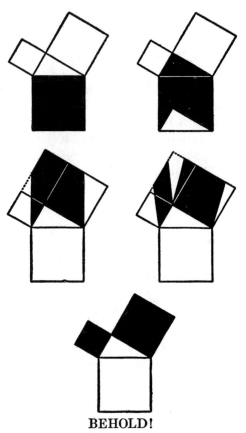

BEHOLD!

Fig. 13.—A Dynamic Proof in a Succession of Five Steps. In the
first step (top left) the square of the hypotenuse is marked in
black. In the second step (top right) a triangle is added on top of
the black square and then an equal triangle is omitted at its base.
Between the second and the third stage (middle left) the form of
the black area does not change; it is just moved upward until its
base line is lifted to the height of its top. In the fourth step
(middle right) the black area is split. Each of the two parts takes
on the form of a parallelogram and is moved sideways with its
bases and heights left unchanged; therefore it preserves its area.
This motion can be continued until the black areas reach their
last phase and become identical with the squares of the legs. This
completes the derivation of the theorem of Pythagoras. It not
only demonstrates that the area of the square on the hypotenuse
equals the sum of the areas of the squares of the legs, but it
shows the actual transformation.

190

worldly goods seems strangely naive if not shocking. Possibly the nameless mathematicians of Babylon did not sign their masterpieces because in their day mathematics was just another daily drudgery like bookkeeping and canal-digging, not worth any sage's claiming. Or they may—as seems probable—simply not have appreciated any significance in what they were doing deeper than that of potential use in their narrow economy. However, they did propose and solve numerous problems of no discernible utility before they succumbed to devastating wars and creeping senility. Pythagoras, with his 'Pythagorean theorem,' seems to have started the tradition of property rights in things of the mind. Today we have such ramified claims as the Smith-Brown-Jones generalization of the Jones-Brown-Smith theorem, neither of which may amount to very much six months after it is printed.

The Pythagorean theorem demanded algebraic restatement before it could be profitably generalized for application to modern mathematics and science. Descartes [7.2] unwittingly provided the germ of this. It is now one of the earliest formulas in any elementary text on analytic geometry. The figure given presently shows up the essential difference between Pythagoras and Descartes. Pythagoras thought in terms of *lengths* and *areas;* so did Descartes, *up to a certain point,* as he used the geometrical form of the Pythagorean theorem in deducing the formula which he used implicitly for the *distance* between two points with given coordinates. But once the formula is derived, no reference to a geometric figure is necessary, as the *visual geometry* of distance has been transformed into an equivalent in terms of *numbers.* Descartes undoubtedly did not appreciate the full significance as we now see it of what he had done. But his successors, including Gauss and Riemann, did, in going far beyond him in the same general direction of rephrasing and generalizing pictorially evident geometric situations into others in which visual intuition survives only in the language in which the generalizations are expressed.

It will be sufficient to see what happens in the first quadrant [7.2] of the figure. By attending to negative coordinates for the other quadrants we readily see that the final conclusion retains its validity for all quadrants.

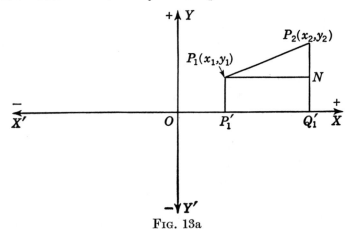

FIG. 13a

The coordinates of any two points P_1, P_2 are (x_1, y_1), (x_2, y_2). In the figure $P_1P'_1$, $P_2Q'_1$ are perpendicular to XOX', and P_1N is perpendicular to $P_2Q'_1$. Hence the respective lengths of $\overline{P_1N}, \overline{NP_2}$ are $x_2 - x_1$, $y_2 - y_1$. By the Pythagorean theorem,

$$\overline{P_1P_2^2} = \overline{P_1N^2} + \overline{NP_2^2};$$

and therefore

$$\overline{P_1P_2^2} = (x_2 - x_1)^2 + (y_2 - y_1)^2;$$

or, finally, *the square of the distance* $\overline{P_1P_2}$ between the points P_1, P_2 whose coordinates are (x_1, y_1), (x_2, y_2) is

$$(x_2 - x_1)^2 + (y_2 - y_1)^2.$$

Remembering what was said about space of many dimensions [7.1], we see that the mere algebra of Descartes' formula for the square of the distance between the two points whose coordinates in Euclidean space are (x_1, y_1), (x_2, y_2) suggests an immediate generalization. Why not imagine a 'Euclidean space' of more than two dimensions, and boldly *define* the *distance* between any two points in it by the immediate algebraic generalization of Descartes' formula? This will justify itself if the generalization satisfies the postulates for 'distance' already given [8.9]. Let us try it for ordinary Euclidean space of *three* dimensions—'solid space' as we commonly conceive it. The square of the distance between the points whose coordinates in such space are (x_1, y_1, z_1) and (x_2, y_2, z_2) is

$$(x_1 - x_2)^2 + (y_1 - y_2)^2 + (z_1 - z_2)^2,$$

and this can be shown to satisfy our previous abstract definition [8.9] of distance. But anyone who cares to sketch the figure will see that we have appealed twice to the Pythagorean theorem in deriving the formula. Thus we have still used our visual intuition. For more than three dimensions such intuition fails. But, by-passing our ancestors who took about 200 years to see the evident thing to do, we follow modern mathematicians and *define* the square of the 'distance' between the 'points' $(x_1, x_2, \ldots, x_n)$, $(y_1, y_2, \ldots, y_n)$ in a 'Euclidean space of n dimensions' to be

$$(x_1 - y_1)^2 + (x_2 - y_2)^2 + \cdots + (x_n - y_n)^2,$$

and then verify that this definition satisfies our previous [8.9] and entirely reasonable postulates for distance. (The verification at one step requires a little skill in manipulating inequalities—a neglected technique in current American courses in school algebra, although it used to be included in so-called advanced algebra.)

I have emphasized that all this applies to a *Euclidean* space, the simplest example of which is a plane. It is sufficient at this stage to state that the *curvature* of such a space is *zero*—that is, the space is 'flat.' But what about 'distances' on a curved surface, say on the surface of the earth where the coordinates of a point are latitude and longitude? The crude Pythagorean theorem is no longer applicable. Latitude and longitude are not measured along straight lines, but along arcs of circles. I refer here to what was said [3.3] about geodesics. The distance between two points on a curved surface is measured on the surface along a *geodesic* joining the two points, and for a sphere the geodesics are arcs of great circles. I need not state the distance formula for neighboring points on a spherical surface, but will remark only that in addition to the *square* terms, as in the Pythagoras-Descartes formula for a *flat* surface, *cross-product* terms occur. This led, but not immediately, to the generalization of 'distance' for any 'space,' 'flat' or 'curved,' in a space of any finite number of dimensions. I shall describe this next.

✤ 10.2 ✤ *From Descartes to Riemann*

In mathematics, as in the arts and literature, the giant of one generation may be a pygmy of the next. The reverse also happens. The farther away we get from some unassuming figure of the past the larger he becomes, and we wonder why his contemporaries failed to recognize his magnitude. Such was Riemann, dead in his prime of a disease that modern medical science might arrest for many years. What Riemann might have accomplished had he lived another ten or twenty years is beyond computation. What he did achieve places him high up in the first rank of mathematicians of all time. Some, including many mathematical and theoretical physicists, have said that in the scope and depth of his thought he surpassed even his master, Gauss. His works fill only one modest volume. His successors say that that volume should be bound in gold.

Riemann's interest for us here is his short but tremendous essay (1854), *The hypotheses which lie at the foundations of geometry.* As we would say in America, this was his dissertation for the Ph.D. degree, although he never took such a degree. He lectured on his essay to an academic audience at which Gauss, as Riemann's professor, presided. Gauss was a man of but few enthusiasms for the mathematical efforts of others. He was within about a year of his death when he heard Riemann's lecture. When it was a question of work of the highest order Gauss could be appreciative, though seldom publicly. He privately told friends that Riemann's lecture was one of the most uplifting experiences of his long life. He himself had started (1827) along the road that Riemann followed to its end, so he, if nobody else in the audience, could understand what his gifted pupil was talking about and foresee its significance for mathematics and the physical sciences.

Few enduring generalizations in mathematics have been mothered by a vacuum. There has usually been a succession of special more or less concrete problems leading up to the grand synthesis by the born master with foresight. It is nothing against a man to have had a mother, so Riemann's great theory need not feel humiliated for having had a grandmother in the practical problems of making accurate maps of the earth's surface. Riemann went far beyond this humble objective and, fully conscious of what he was doing (as will appear shortly), undertook the drafting of a map of the physical universe. The problem of drawing terrestrial maps for the demands of taxation, navigation, and commerce is as old as ancient Egypt, possibly older. It began to become of deep mathematical significance when human beings, at last accepting the testimony of astronomers and navigators, admitted that the earth is round, not flat as prescribed by Holy Writ. Directly inspired by the work of Lagrange, who in turn had been motivated by a long succession of cartographers and navigators, Gauss in 1827 published his classic on the geometry of curved surfaces. This probably was Riemann's point of departure. We must skip the intermediate steps and come directly to Riemann's generalization which now, since general relativity has familiarized us with it, seems like another of those things that should have been imagined long before it was. But if Gauss in his prime overlooked this, what chance have ordinary mathematicians of our own day in seeing what may be of fundamental importance to their successors a decade or a century hence?

Riemann took as the underlying idea of his geometry a 'manifold' of n dimensions. The 'elements' of his manifold were not necessarily numbers, and in that he anticipated to a certain extent those abstract geometries of the twentieth century which we have noted [8.9]. But for applications to physics it is sufficient to think of the elements of a manifold as ordered sets of real numbers in the manner of Descartes [7.2]. As the concept of such a manifold is basic for applications of Riemann's ideas to current physics, I must describe it in full, possibly at the cost of elaborating some things that have already been intimated.

The letters $x_1, x_2, \ldots, x_n, y_1, y_2, \ldots, y_n$ shall denote real numbers [4.2]; n is a positive integer which may be as large as we please. By enclosing $x_1, x_2, \ldots, x_n$ in parentheses, thus $(x_1, x_2, \ldots, x_n)$, we mean that the numbers $x_1, x_2, \ldots, x_n$ are arranged in the definite order x_1 first, x_2 second, $\ldots, x_n$ nth. If a name for this ordered set of

n numbers is wanted, it may be called an *entuple*, or an *n-tuple*. This is merely the generalization of Descartes' invention of coordinates from two or three dimensions to n.

Two entuples $(x_1, x_2, \ldots, x_n)$, $(y_1, y_2, \ldots, y_n)$ are said to be *equal* when, and only when, $x_1 = y_1$, $x_2 = y_2$, $\ldots, x_n = y_n$. If these entuples are equal we write

$$(x_1, x_2, \ldots, x_n) = (y_1, y_2, \ldots, y_n).$$

With the given numbers 1, 2, 4, for example, we can construct precisely 6 distinct triples, $(1, 2, 4)$, $(1, 4, 2)$, $(2, 4, 1)$, $(2, 1, 4)$, $(4, 1, 2)$, $(4, 2, 1)$; with 1, 2, 2, only 3 distinct triples, $(1, 2, 2)$, $(2, 1, 2)$, $(2, 2, 1)$. Equality as thus defined is an example of an equivalence relation [5.3].

In the typical entuple $(x_1, x_2, \ldots, x_n)$ we now let $x_1, x_2, \ldots, x_n$ vary, each over some prescribed range of real numbers. The set of all the distinct entuples generated from $(x_1, x_2, \ldots, x_n)$ in this way is called a *number-manifold*, or simply a *manifold, of n dimensions.* I have already roughly described manifolds [8.1] in connection with Klein's geometry. It is necessary now to make the notion more precise. If $x_1, x_2, \ldots, x_n$ in varying take only discrete [4.2] values (for instance each might traverse the set of all integers), the manifold is called *discrete;* if $x_1, x_2, \ldots, x_n$ vary continuously [4.2] the manifold is called *continuous.* Thus if each of $x_1, x_2, \ldots, x_n$ can range over the real number system [4.2], a continuous manifold of n dimensions is generated. There is an intermediate possibility in which some of the x's vary continuously and the rest discretely. But as such a manifold can always be split into a continuous and a discrete manifold, there is no need to discuss it independently.

The preceding definition of a manifold of n dimensions is sufficiently general for most scientific purposes. It is not the most general definition conceivable or useful, as is plain from our previous notice [8.9] of abstract spaces.

So far I have defined entuples, their equality, and manifolds. Two manifolds are said to be *equal* if each entuple which occurs in either one occurs also in the other. We thus have the notion of distinct manifolds. What can be done with a particular manifold when we have it? We can *geometrize* it. With one eye on what Descartes did for his pairs and triples, we can talk about the entuples in a given manifold in the language of geometry. This may seem like cheating after having turned our backs on intuitive geometry, and perhaps it is. Possibly it is merely lack of imagination.

The first geometrical concept to be carried over to manifolds is that of distance, and this is done for *Euclidean* manifolds of n dimensions as already explained [10.1].

The next is also suggested by analogy with geometry, but not Euclid's geometry. We noted [10.1] that the crude Pythagorean theorem requires modification for the surface of a sphere where coordinate distances are measured along arcs of great circles. To take account of reasonable possibilities, the generalization of the distance formula to any number-manifold must provide for variability *in the neighborhood of a point*. The appropriate mathematics is called *differential geometry*, of which there are many kinds. For example, if in space of three dimensions a surface has humps or spines on it, we cannot prescribe a 'distance formula' which will be valid for *any* two points a *finite* (non-zero) distance apart. So we must define distance for two *sufficiently close* points on the surface, and if we can formulate our definition so that it is valid for *any* pair of such points, we shall have defined 'distance' for the entire surface. All this is carried up at once to manifolds of n

dimensions. Our manifolds now are not necessarily Euclidean, or 'flat,' so the previous formula [10.1] for distance is no longer sufficient. We proceed as follows.

We start from the generalized Pythagorean formula [10.1] for the square of the distance between two entuples in a Euclidean manifold of n dimensions,

$$s^2 = (x_1 - y_1)^2 + (x_2 - y_2)^2 + \cdots + (x_n - y_n)^2.$$

We re-express it for 'neighboring' entuples, examine what characteristics the new formula has, and finally *generalize* these characteristics so that the generalized formula shall apply *in the special case of a two-dimensional manifold* ($n = 2$) *to any surface in ordinary three-dimensional space.* The last is accomplished by inspecting the distance formulas for spheres, ellipsoids, and other surfaces with which we are familiar. The relevant formulas for the latter were first discovered by brute force.

Generalizing our experience for $n = 2,3$, we then express the fact that $(y_1, y_2, \ldots, y_n)$ *is in the neighborhood* of $(x_1, x_2, \ldots, x_n)$. This means only that we shall take the numbers $y_1, y_2, \ldots, y_n$ so close to $x_1, x_2, \ldots, x_n$ respectively that *all powers of the differences* $x_1 - y_1$, $x_2 - y_2$, $\ldots, x_n - y_n$ *higher than the second can be neglected in all our calculations.* This, of course, is only a definition. If greater refinement were required we might keep all powers up to the third, or fourth, and so on, of the *small* differences $x_1 - y_1, x_2 - y_2, \ldots, x_n - y_n$. But what has been defined (retention of only first and second powers) has been found sufficient for scientific purposes.

Another way, perhaps simpler, of saying that $(y_1, y_2, \ldots, y_n)$ is in the neighborhood of $(x_1, x_2, \ldots, x_n)$ is this:

$$y_1 = x_1 + d_1, y_2 = x_2 + d_2, \ldots, y_n = x_n + d_n,$$

where d_1, d_2, . . . , d_n are numbers (positive or negative) so small that, for our purposes, we can discard all powers of them higher than the second. This gives us, for the required formula,

$$s^2 = d_1^2 + d_2^2 + \cdots + d_n^2,$$

where s, d_1, d_2, . . . , d_n are all so small that their powers higher than the second can be neglected.

Here we introduce a most convenient shorthand for expressing the fact that y_1 is close to x_1, y_2 close to x_2, and so on, in the sense just explained. This will be familiar to anyone who has glanced through a book on relativity, and it will make our formulas look more orthodox. The *small* number d_1 is written dx_1, since it refers to (is 'near to') x_1; d_2 is written dx_2, and so on. Also to express the fact that s is now correspondingly small, we write ds. The d may be read 'differential of,' but it is usual simply to read dx by pronouncing the letters d and x; dx is called the *differential* of x, and similarly for ds and the rest. The preceding formula for the distance from $(x_1, x_2, . . . , x_n)$ to any *neighboring* point is now

$$ds^2 = dx_1^2 + dx_2^2 + \cdots + dx_n^2,$$

since $y_1 - x_1 = d_1$, . . . , $y_n - x_n = d_n$. More properly we should write $(ds)^2$, $(dx_1)^2$, . . . instead of ds^2, dx^2, . . . , but custom and convenience sanction the shorter way. All this will be familiar to anyone acquainted with 'differentials' as in the calculus, but there is no need here to introduce essentially extraneous difficulties.

The most significant algebraic feature of this formula is evident: it is of the *second degree in the differentials ds*, dx_1, . . . , dx_n. The above expression for ds^2 is called a *quadratic* ('second degree') *differential form*.

To proceed at once to the generalization, we replace this

special quadratic differential form by the general *quadratic differential form*, which I shall write out in full for a continuous manifold of four dimensions ($n = 4$).

$$ds^2 = g_{11}dx_1^2 + g_{22}dx_2^2 + g_{33}dx_3^2 + g_{44}dx_4^2 + 2g_{12}dx_1dx_2$$
$$+ 2g_{13}dx_1dx_3 + 2g_{14}dx_1dx_4 + 2g_{24}dx_2dx_4 + 2g_{34}dx_3dx_4.$$

The form is called *quadratic* because all terms in it, such as $g_{11}dx_1^2$, $2g_{14}dx_1dx_4$, are of the *second degree* in the differentials $dx_1, \ldots, dx_4$; it is *general* because all such terms have been included. All this will be familiar to those who have inspected books on general relativity. The $g_{11}, \ldots, g_{34}$ denote any functions [4.2] of the variables x_1, x_2, x_3, x_4. For any n the formula for ds^2, the square of the distance between neighboring n-tuples in the manifold of n dimensions, is precisely similar. The coefficients 2 for the cross-product terms are inessential and could be suppressed except for reasons of algebraic convenience in the derivation and statement of formulas.

Measurements in a manifold are based upon distance-formulas. So we say that the 'metric,' or theory of measurement, for the manifold is given by the quadratic differential form for the ds^2 appertaining to the manifold. The mathematics of this metric, also the metric properties of the manifold, are completely specified when the functions $g_{11}, \ldots, g_{34}$ are given, and so for any n. As these functions may be chosen at will, we have the foundation for an infinity of different 'geometries.' Mathematicians investigate these en masse unless, as in the case of general relativity, there is a special reason for studying a particular set of g's intensively.

As Riemann was the first to propose a geometry of n dimensions with a *general* quadratic differential form as its metric, such geometry is named after him. In his great essay he envisaged possibilities other than the quadratic form universal now in what is usually called *Riemannian geometry*. Each of the further possibilities has the characteristic properties of a distance [8.9], but none has been developed, possibly because physics has not yet had a use for any of them. After all, as even the purest of pure geometers must admit, it was only when general relativity (1916) popularized Riemannian geometry that geometers began to write innumerable papers on it. Of course it had been worked at sporadically before relativity showed its scientific importance, but the spate did not begin till 1917.

One of the things defined in Riemannian geometry in analogy with common geometry must be mentioned, as it is of importance in physics. This is *curvature*. Attempts to visualize curvature in a Riemannian manifold of more than two dimensions are profitless. Curvature in the four-dimensional Riemannian manifold on which general relativity is based is a purely mathematical concept which was first devised by formal generalization from the like in the visualizable case of two dimensions. I do not mean to imply that it is impossible to draw diagrams on a flat sheet of paper actually exhibiting the curvature of a four-dimensional manifold, for this can easily be done. But all such pictorial representations are hopelessly confusing and none of them can show us anything that is not much more simply obtained by analysis.

Roughly, Riemann's generalization of curvature came about as follows. Suppose we wish to give a mathematical description of our intuitive notion of what is meant by one portion of a surface (say a streamlined fender) being 'more curved' or 'less curved' than another. Draw a small circuit round the point at which the curvature is to be measured. At each point of this circuit, imagine a perpendicular drawn to the surface—that is, perpendicular to the tangent plane at the point. At any convenient place put a sphere whose radius is one unit of length. Through the center of this sphere imagine lines drawn parallel to the perpendiculars just described. These parallels will cut out a curve on the surface of the sphere. The small area enclosed by this curve is a kind of image of the humpiness of the small area enclosed by the curve on the original surface, and is a measure of the total curvature of that small area. There are other ways of estimating curvature, but this is sufficient here.

Now, the ds^2 for an ordinary surface (two-dimensional manifold) is given by

$$ds^2 = g_{11}dx_1^2 + 2g_{12}dx_1dx_2 + g_{22}dx_2^2,$$

and it is found that any of the formulas giving the measure of curvature can be expressed wholly in terms of g_{11}, g_{12}, g_{22}. Proceeding exactly as in the step from a 'space' (manifold) of two dimensions to one of n—namely, by formal generalization—we can construct a formula of the same 'curvature' sort for the manifold of n dimensions. This formula will involve only the g's defining the metric of the manifold.

That this process of formal mathematical generalization should lead to anything of physical significance seems like a miracle, but it is a fact. The 'geometry'—the abstract mathematics—of Riemannian manifolds and their curvature was highly elaborated a full generation before Einstein found a physical use for any of it in his theory of relativity. Without this geometry, relativity—at least the general theory, including the theory of gravitation—might have been delayed perhaps for a century. It was sheer luck that a man with Einstein's penetrating physical insight appeared at the dramatic moment. This may suggest to some ambitious aspirant to scientific fame that similar conquests are no doubt still to be financed with the unexhausted and inexhaustible gold of pure mathematics.

❖ *10.3* ❖ *From Riemann to Einstein*

Riemann concluded his essay with the following remarks. They contain incidentally a prophecy which Einstein was to validate. I quote from Clifford's translation (1873).

> Either therefore the reality which underlies space must form a discrete manifold, or we must seek the ground of its metric relations outside it, in binding forces which act upon it.
> The answer to these questions can only be got by starting from the conception of phenomena which has hitherto been justified by experience, and which Newton assumed as a foundation, and by making in this conception the successive changes required by facts which it cannot explain.

He then says that researches like his own, starting from general notions,

> can be useful in preventing this work from becoming hampered by too narrow views, and the progress of knowledge of the interdependence of things from being checked by irrational prejudices.
> This leads us into the domain of another science, physics, into which the object of this work does not allow us to go today.

Had he lived, Clifford (who died at thirty-four of the same disease as Riemann) might have been a worthy successor to Riemann. For in 1870 he published a short note *On the space-theory of matter*, in which he also prophesied in the manner of Riemann and in more detail. As what he said is a remarkable foreshadowing of some aspects of twentieth-century physics, I shall quote him.

> Riemann has shown that as there are different kinds of lines and surfaces, so there are different kinds of space of three dimensions; and that we can only find out by experience to which of these kinds the space in which we live belongs. In particular, the

axioms of plane geometry are true within the limits of experiment on the surface of a plane sheet of paper, and yet we know that the sheet is really covered with a number of small ridges and furrows, upon which (the total curvature being not zero) these axioms are not true.

Similarly, he says,

although the axioms of solid geometry are true to within the limits of experiment for finite regions of our space, yet we have reason to conclude that they are true for very small portions; and if any help can be got thereby for the explanation of physical phenomena, we may have reason to conclude that they are not true for very small portions of space. [Gauss had some of these ideas. But Clifford knew nothing of them as they were not published till years after both he and Gauss were dead.]

I wish here to indicate a manner in which these speculations may be applied to the investigation of physical phenomena. I hold, in fact,

(1) That small portions of space *are* in fact of a nature analogous to little hills on a surface which is on the average flat; namely, that the ordinary laws of geometry are not valid in them.

(2) That this property of being curved or distorted is continually being passed on from one portion of space to another after the manner of a wave.

(3) That this variation of curvature is what really happens in that phenomenon which we call the *motion of matter*, whether ponderable or ethereal. [He was writing in 1870, while the ether was still in fashion.]

(4) That in the physical world nothing else takes place but this variation, subject (possibly) to the law of continuity.

As for (1), in relativity the degree of curvature of space depends upon the amount of matter present. In (2), instead of 'waves' we might imagine the transformation of matter into energy, and vice versa. But perhaps we had better not.

The two main connecting links uniting Riemannian geometry with relativity are the *space-time interval* between *neighboring point-events* in a four-dimensional space, and Einstein's hypothesis that this interval is *invariant* for continuous [4.2] transformations of the four coordinates of events. The justification for the hypothesis is the pragmatic one, as C. S. (S.) Peirce (1839–1914) and W. James (1842–1910) might have said, that it works. I shall explain first by an example what a space-time interval is. The example is simple to the point of platitute; the rest is not. In fact I know of no argument, mathematical, scientific, or philosophical, which builds up this cardinal physical concept of a space-time interval from anything simpler or closer to intuition, nor of any conclusion as simply arrived at that is so far from intuition.

About the simplest physical event imaginable is a collision of two particles. To avoid unnecessary complications I shall consider particles—mere 'mass-points'—instead of sizable bodies. Analyzing the collision in order to translate its salient characteristics into mathematical symbolism, I note that it occurs at a definite place at a definite time. This is enough.

To fix the place and time of the collision we must require some standard space-time reference system, in short a system of coordinates in space and time. For the space coordinates we may choose three mutually perpendicular axes anywhere, say the lines in which the north wall of a room, the east wall, and the floor intersect. Say the coordinates of the point of collision with respect to these axes are (x, y, z). To fix the time t of the collision, we read the time (in seconds, say) on a standard clock. Thus if the collision occurred at 3.1415 seconds past 12 o'clock, we should take $t = 3.1415$, reckoning 12 o'clock as the 'zero' or *origin* of time. Wherever and whenever the collision occurred, it could be located in space-time by measuring its (x, y, z), say in inches, and its t, say in seconds. The quadruple

(x, y, z, t) would then be the *space-time coordinates* of the collision. The collision is called a *point-event*.

Next, consider two collisions, not necessarily at either the same time or the same place. Proceeding as before we get the space-time coordinates (x_1, y_1, z_1, t_1), (x_2, y_2, z_2, t_2) of these two point-events. These events are separated by a *space-time interval*. How shall this interval be measured? I cannot repeat the arguments which led up to the conclusion, as they belong to physics rather than to mathematics, but I may briefly recall the simple and profound observation on which they were based.

The numbers x,y,z,t in the coordinates (x, y, z, t) of a point-event mean nothing physically unless some procedure is specified for determining them experimentally by performable measurements. But all laboratory measurements depend ultimately upon comparing coincidences on standard measuring devices. Thus to read a temperature we observe what point the top of the mercury column is opposite on the graduated scale etched on the thermometer. This is a matter of eyesight, which in its turn depends upon the reception of light signals. Light, however, is assumed to travel in free space (space devoid of matter, or approximately so) with a constant velocity; velocity involves both 'space,' or 'distance,' in the ordinary sense, and 'time.' When fully elaborated this remark leads to the conclusion that actual measurements involve both space and time and that it is impossible to separate the two in consistent and empirically meaningful descriptions of physical experience.

The statement that the velocity of light is constant in a vacuum was one of the basic postulates of the special theory of relativity (1905). This assumption was justified by experience, also by physical arguments into which I shall not go. Let c denote this constant velocity. Then in the sense described in the preceding section, the space-time interval

ds between neighboring point-events was taken (*an assumption*) to be

$$ds^2 = dx^2 + dy^2 + dz^2 - c^2\,dt^2$$

in the special theory of relativity. Again I cannot enter into the physical arguments on which this assumption was based. Our interest here is in the mathematics that followed. The formula for ds^2 provided the space-time of special relativity with its metric.

Comparing the above formula for ds^2 with that in [10.1] for the distance between neighboring points in a space of four dimensions (three of 'space,' one of 'time'), we note a shocking lack of generality. Why should all the g's in the general formula be constants? The question of course is meaningless, except possibly to mathematicians who are seldom content with anything short of the completest generality attainable in any given direction. Here the direction is that of quadratic differential forms in four variables. Why not assume as the metric of physical space-time the general form and develop the consequences mathematically? This is what Einstein finally did, but for no such reason as that just suggested. His motives appear to have been mostly scientific, partly aesthetic, and perhaps a little mathematical. Pure mathematicians criticized the special theory because it clashed with the *general* notion of invariance. They were right. This is the point of interest here.

Suppose two observers are describing the same physical event, say a collision of two particles. Each will describe the event with respect to *his own* space-time coordinate system. Assuming that *the event is unaltered* by either observer's way of recording and describing it, we see that the mathematical description must retain its form when *either* observer describes the event in terms of *the other* observer's space-time coordinate system. That is, the mathematical equations of a natural, nonhuman event must be invariant for all space-time coordinate systems in terms of which the event is described. 'Must' is probably too strong; it is safer to take the foregoing assertion as a postulate. Indeed, in the quantum theory the opposite is assumed to be significant for certain small-scale phenomena; namely, the very act of exploration by physical apparatus becomes a part of any observation, and it is impossible to disengage the experimenter from his experiment. This possibility was quite obscurely explored by N. Bohr (contemporary), particularly for its significance for experimentation in biology. But in relativity it is assumed that events are invariant in the sense explained.

If now one observer uses another's coordinate system for constructing his equations, he must have some means of expressing his own coordinates in terms of the other man's. In mathematical language, either observer requires a set of equations expressing his coordinates in terms of the other's. This will express the x,y,z,t of the first's in terms of the x',y',z',t' of the second's, and conversely. When the first man replaces his x,y,z,t by their expressions in terms of x',y',z',t', his equations in x,y,z,t must reduce to new equations in x',y',z',t' of exactly the same form as his original equations in x,y,z,t. Similarly for the second man. This amounts to demanding that the equations describing natural phenomena be invariant under all transformations of the space-time coordinates. Equations which do not meet this condition of invariance are tainted with peculiarities due to the observer's reference system and are not intrinsic in nature.

It therefore becomes a capital problem to reformulate the equations of classical mathematical physics in invariant form. These are *differential equations* in a technical sense which will be explained later [15.4] and which need not detain us now. Such equations express relations between rates of change—for example, velocities and accelerations.

The reformulation was easily accomplished in special relativity where, however, invariance was demanded only in the drastically restricted form that the equations were to be invariant for all transformations expressing the fact that one observer was moving with *uniform* (constant) velocity relatively to another. Why should such extremely special motions be singled out for exceptional favor? That they were, was one of the defects of the special theory of relativity. Another was that certain predictions made by that theory were contradicted by observations. The last was fatal. It took ten years of intense thought on Einstein's part to overcome these difficulties.

The difficulties were serious enough. At first sight they might have appeared insuperable. For if we demand that our descriptions of natural phenomena shall be unchanged when the reference system is *accelerated*, we are asking that the effects of certain *forces* be eliminated. This flies in the face of all classical (Newtonian) dynamics.

The mathematical machinery for constructing the desired universally invariant equations had been fully elaborated for about twenty years before Einstein attacked the problem. Originated essentially by Riemann before 1860, in a posthumously published memoir on heat conduction prepared in competition for a prize offered by the French Academy of Sciences (what forgotten competitor won the prize?), this machinery, now called *tensor analysis*, was brought to a high degree of utility in 1887 by M. M. G. Ricci (1835–1925) and T. Levi-Civita (1873–1941).

Tensor analysis, or the *tensor calculus*, was not one of the tools in the kit of the well-equipped mathematical physicist of 1905, or even of 1916 when general relativity made it indispensable. Luckily Einstein fell in with a geometer, M. Grossmann (1878–1936) of Zurich, who was thoroughly familiar with the work of Ricci and Levi-Civita. Grossmann taught Einstein the tensor calculus (he had a hard time mastering it), and Einstein did the rest. It was one of the few ideal collaborations in the history of science. Without the other partner it is doubtful (at least to mathematicians) that either could have accomplished the best that was in him. Newton invented the differential calculus [15.1] to aid him in his analysis of motion, and it is possible that Einstein might have been driven to invent the tensor calculus had not the mathematicians had it ready for him when he needed it. But the complexities of the problem he faced make such a supposition seem unlikely. As an indication of the greatly accelerated progress of both science and scientific education since the 1920s, the tensor calculus that cost Einstein an effort to master is now a regular part of an undergraduate course in the better technical schools. The subject has been so thoroughly emulsified that even an eighteen-year-old can swallow it without regurgitating. But this does not prove that either his brain or his stomach is stronger than Einstein's was.

Mathematics by itself has seldom got very far in the exploration of nature, as is attested by the numerous attempts of pure mathematicians of the past and present to solve the universe with pencil and paper. There are exceptions, of course—Newton and Einstein. Before predictions can be made, hypotheses connecting mathematics with the observable universe must be laid down. We cannot get invariants out of nature unless we put at least one into nature in the first place. Guided by analogy with Newtonian dynamics and the special theory of relativity, Einstein *postulated* that ds^2, where ds is the space-time interval separating neighboring point-events in the four-dimensional space-time continuum of physics, is an invariant. That is, this ds^2 as measured by all observers will be the same. He also *postulated* that this ds^2 will be given by the general formula in [10.2]. The last was a direct generalization from classical mechanics and special relativity.

As already indicated the ultimate justification for this abstruse assumption—the invariance of ds^2—is pragmatic.

Again guided by analogy with classical mechanics, particularly Newton's first law of motion [13.3], namely, that every body will continue in its state of rest or of uniform motion in *a straight line* unless it is compelled to change that state by an 'impressed force,' Einstein *postulated* that the path of a free particle in the space-time continuum of general relativity is a geodesic [3.3] of the space. Further reasonable assumptions generalizing the Newtonian situation provided adequate material for the tensor calculus to take hold of. Two of these may be mentioned in passing. The second seems reasonable enough; the first is somewhat shocking to common sense, but that is only because a platitude is clothed in an unfamiliar mathematical dress.

For physical reasons it was *assumed* that the path of light in the space-time continuum [4.2] is such that $ds^2 = 0$ along the path.

The second assumption is closer to common experience. It is easily visualized through Einstein's classic fable of the elevator. A man falling freely feels no tug of the 'force of gravity' on his body. He has no 'weight,' nor has anything in his pockets. Weight in Newtonian mechanics is due to the force of gravity. Other bodies in the world besides that of the falling man will have weight. For instance if the man falls down an elevator shaft, the heavy cables he tries to clutch on his way down have not lost their weight merely because he has. In the same way if the cables sustaining the elevator break, the elevator loses its weight as it is accelerated downward. So do all passengers in the elevator, until they hit the bottom of the shaft. The 'force of gravity' has been abolished for them by an equivalent downward acceleration of their coordinate system (the walls and floor of the elevator). An anecdote, which may be apocryphal, emphasizes the point. Einstein saw a carpenter fall off a scaffolding. Luckily the man landed on a pile of straw and was unhurt. Einstein rushed over to him and asked whether he had felt anything pulling him down as he fell. The man said "No." Any reader who doubts the man's assertion may check it by jumping off the roof.

'Forces,' in particular the 'force of gravitation,' thus appear as equivalent in their effects to accelerations of the observer's coordinate system. It was *postulated* then that any gravitational field *in a sufficiently small region* of space-time is equivalent to a transformation of the coordinates in that region. 'Forces' do not appear in this sort of physics.

One further *assumption* may be mentioned: at a great distance from all matter, the ds^2 of general relativity (see [9.1]) degenerates to that of special relativity. In the absence of matter, space, technically, is 'flat.'

We noted the fundamental postulate of invariance for the mathematical equations of physical phenomena. We also remarked the assumption that ds^2 is an invariant. These are physical assumptions, imposed on the geometry of the space-time manifold whose metric is a quadratic differential form [9.1]. From all this it seems reasonable to suppose conversely that *any* equations in invariant form, constructed from the g's [10.1] in the ds^2 appropriate to a specific problem, will be the mathematical expression of some physical phenomenon. Similarly for any mathematical invariant constructed from the g's. One invariant of this sort gives the curvature of the manifold. When interpreted physically this invariant gave the theory of the gravitational field from which Einstein made his most striking predictions. Before recalling these I may state the gist of the mathematical method.

The tensor calculus is the proper machinery for discovering those differential equations [15.4] which remain unchanged in form (invariant; *covariant* is the usual technical term) for all (analytic [4.2]) transformations of their variables. The variables in these equations are the coordinates of a 'point' (or quadruple) in the four-dimensional space-time manifold. The metric of this manifold characterizes the theory of measurement in the manifold. The Riemannian geometry [10.2] of the manifold is developed by the tensor calculus and is reinterpreted in terms of physics. Throughout the entire process the language of geometry—some would say geometric intuition—is the guide. The abstract geometry suggests many things to do. These are tried, checked against physical intuition or experience, and developed if they are reasonably close to what is found.

This powerful method of physical geometry is not omnipotent. Problems of apparent simplicity but of the greatest scientific interest, whose solutions we should like to have, are at present too hard for the mathematicians to solve. Of such we may exhibit this innocent specimen: if two particles are let loose in a gravitational field, how will they behave? In Newtonian gravitation this is Kepler's problem, described in Chapter 13, now solved in a page or two. In Einstein's gravitation a solution has been sought in vain since 1922. Several have been proposed, but none is even moderately satisfactory. The most promising attempt was that (1937) by Levi-Civita. It predicted an effect on the motions of binary stars that should have been observable. However, before any observations were attempted, H. P. Robertson (1903–) found (1938) a mistake in the elaborate calculations, causing the abandonment of the whole. Here is a genuine problem (not an artificially manufactured one) for some ambitious and gifted young mathematician to attack.

Before passing to an account of the verified predictions of general relativity, I add a more technical note to indicate why the tensor calculus of Ricci and Levi-Civita should have worked as it did. Those not interested may skip this and pass at once to the verified predictions.

A general (analytic [4.2]) transformation on the variables in a *tensor* (as in the theory of Ricci and Levi-Civita) transforms the tensor into another whose components are *linear homogeneous functions* of the components of the original tensor. A tensor, like an ordinary vector, vanishes (is equal to zero) if and only if each of its components vanishes. A transformation of the kind stated is, geometrically, a general (analytic [4.2]) transformation of coordinates, when the variables in the tensor are interpreted as coordinates in a space of the appropriate number of dimensions. Hence if a tensor vanishes in one system of coordinates, it vanishes in all; the decisive detail is the homogeneity. This is equivalent to saying that if a system of equations is expressible as the vanishing of a tensor, then the system will be invariant under all (analytic) transformations of the variables in the system. But this is precisely the condition imposed by one of the postulates of general relativity on a system of equations, if the system is to be an admissible mathematical formulation of an observable sequence of events in physics or cosmology.

I shall now briefly state the verified predictions of the theory of general relativity. A fair share of the credit for these should go to the mathematics without which none would have been possible. Three of these predictions led to the discovery of new phenomena, not even suspected until the predictions were made. What is equally remarkable, the predictions were quantitative as well as qualitative. They said, "Look for such and such an effect, measure it, and you will find a certain number expressing the amount."

According to the Newtonian law of gravitation [13.3] the orbit of a single (unperturbed) planet is an ellipse with the sun at one focus. In Einstein's theory the orbit is almost an ellipse, but the path does not close after one revolution of the planet. At each revolution the orbit advances by a small amount in the direction of motion of the planet. Although small, the advance is easily measurable in a century. Einstein's theory predicted for the planet Mercury an advance of 43 seconds of arc per century *in addition to* the advance of 532 seconds per century due to the perturbations of the other planets as calculated on the Newtonian theory. The *observed* advance is 574 seconds per century. Einstein's correction is only 1 second off the observed discrepancy, well within the limits of observation.

A remarkable feature of this prediction was that it followed directly from the theory without the addition of special hypotheses. Before Einstein many others had tinkered with the Newtonian law in *ad hoc* attempts to explain the discrepancy—there is no difficulty in modifying the Newtonian law to accommodate *a single* discrepancy. So conspicuous was this blemish on the Newtonian theory of the solar system that the great mathematical astronomer U. J. J. Leverrier (1811–1877), whom we shall meet later in the company of the planet Neptune, confident of repeating his success with Neptune, calculated the possibility that another planet would be found between Mercury and the sun. Before he began his search, the hypothetical planet had even been named—Vulcan. Until June, 1949, only a few stray asteroids that might have been Vulcan had been observed. One of the early photographs taken in 1949, with the 48-inch Schmidt camera-telescope at the Palomar

Mountain Observatory, disclosed a miserable little asteroid nine-tenths of a mile in diameter where Vulcan might have been but was not. They should have called it Periphetes. This clod was by no means the first candidate for the honor of being Vulcan. Its orbit ruled it out as the dubious planet. So Leverrier seems to have been too confident.

The second prediction concerned light. If light has mass, it should be deflected on passing a massive body, say the sun. The amount of the deflection can be calculated by Newton's theory on the assumption that light is corpuscular. Without making any hypothesis as to the nature of light, beyond that already mentioned about the *ds* for a light track, Einstein's general theory predicts a deflection which, for light grazing the sun, is *twice* the Newtonian prediction. Precise observations at several eclipses have confirmed the prediction made in accordance with general relativity.

The third prediction also concerned light. According to the general theory of relativity, the spectral lines in light issuing from a massive body should be shifted toward the red end of the spectrum by an amount which can be calculated (depending on the mass of the body). This was confirmed for the sun and, more brilliantly, for the massive little companion star to Sirius.

There is a fourth consequence of the theory (plus some additional hypotheses) concerning the recession of the spiral nebulae, but as this is still (1950) under investigation I shall pass it. The other three are enough for any mathematical-physical theory to have predicted. Even if the theory should be discarded next week it will have played its part in leading science to unforeseen vistas opening out on the observable universe. The omitted fourth consequence, for example, has inspired much research in observational astronomy. There seems to be a hope that the 200-inch Hale telescope on

Palomar Mountain, 'unveiled' to an invited public on June 3, 1948, may give some information regarding the hypothesis of an 'expanding universe.' This imaginative hypothesis was an offspring of relativistic cosmology. It has been described so often and so well in the press and popular books on astronomy that there is no need here to say what it is about.

It is always interesting, if sometimes deflating, to hear what others have to say of the things that we think make life more endurable. At a dinner someone mentioned science during a lull in the wrangle between the guests whom the wily host had trapped into expressing their political opinions. Soon the differences of opinion on matters scientific became almost as chaotic as the previous political anarchy. In the thick of the fight an ex-judge hurled this grenade at an eminent lawyer who many times had fought his battles right up to the Supreme Court of the United States:

"Why, you don't even know that the Newtonian law of universal gravitation has been repealed and is definitely out."

"Has it? And is it?" the lawyer enquired innocently. "Since when?"

"Oh, since 1916."

"Funny! I've never missed it."

It takes an atom bomb or a potential plague to convince some of our good citizens that abstract mathematics, theoretical physics, and unobtrusive puttering in biological laboratories are not of interest only to dreaming professors. Mathematicians, physicists, biologists, and bacteriologists seldom dream nowadays. When they nod and visualize what use the saviors of this, that, or the other creed would make of their discoveries they have nightmares.

As a fitting tailpiece to this chapter I report the query of the man who, of all living astronomers, has enlarged our knowledge of the universe almost beyond human conception. Assured by some rosy social optimist that 'science' would find an answer to the urgent problems of survival of the human race, including an adequate food supply, he said, "Yes, Science will find the answers. And what will the human race do with them?"

Chapter 11

THE QUEEN OF MATHEMATICS

✤ 11.1 ✤ An Unruly Domain

The Servant having done most of the talking in the preceding chapter, it is now the Queen's turn. She will have something to say about her favorite domain, numbers.

Gauss, I recall, crowned arithmetic the Queen of Mathematics. Gauss lived from 1777 to 1855, and to his profound inventiveness is due more than one deep and broad river of mathematical progress during the nineteenth and twentieth centuries. In the higher arithmetic, or the theory of numbers, his work is as vital as it was in 1801, when he published his *Disquisitiones arithmeticae*. Some of his innovations in that masterpiece, for example the concept of congruence [5.3], have had a far wider significance than even he may have forseen. The like holds for his contributions to geometry, which Riemann [10.3] was to generalize in preparation for the physics of the twentieth century. He also made outstanding contributions to the science of his time, notably to electromagnetism and astronomy. His opinions therefore are respected by all mathematicians and by some scientists.

Arithmetic to Gauss, as to the Greeks, was primarily the study of the properties of the whole numbers. The Greeks, it may be remembered, used a different word for calculation and its applications to trade. For this practical kind of arithmetic the aristocratic, slave-owning Greeks seem to have had a sort of contempt. They called it *logistica*, a name which survives in the logistics of one modern school in the logic and foundations of mathematics. However, without this despised drudge of the Queen even the most aloof of the great Greek astronomers could not have gone very far with his "System of the World"—in Laplace's phrase.

In arithmetic, as in all fields of mathematics since 1920, discovery went wide and far. But there was one most significant difference between this advance and the others. Geometry, analysis, and algebra each acquired one or more vantage points from which to survey its whole domain; arithmetic did not.

The Greeks left no problem in geometry which the moderns have failed to dispose of. Faced by some of the trifles, like that of 'perfect numbers,' which the Greeks left in arithmetic we are still baffled. For instance, give a rule for finding all those numbers which, like 6, are the sums of all their divisors less than themselves, $6 = 1 + 2 + 3$, and prove or disprove that no odd number has this property. Such a number is called *perfect;* the next perfect number after 6 is 28. We shall meet these numbers again. To say that arithmetic is mistress of its own domain when it cannot subdue a childish thing like this is undeserved flattery.

The theory of numbers is the last great uncivilized continent of mathematics. It is split up into innumerable countries, fertile enough in themselves, but all more or less indifferent to one another's welfare and without a vestige of a central, intelligent government. If any young Alexander is weeping for a new world to conquer, it lies before him. Arithmetic has not yet had its Descartes, to say nothing of its Newton.

Lest this estimate seem unduly pessimistic, let us not forget that in each of the several countries of arithmetic there was remarkable progress since the time of Gauss, and especially since 1914, when modern analysis [4.2] was applied to problems in the theory of numbers that had withstood the strongest efforts of Gauss's successors for over a century. Indeed, two or three of the splendid things done are comparable to anything in geometry, with this qualification, however: no one advance affected the whole course of development. This possibly is due to the very nature of the subject.

Among the notable advances is that which revealed one source of some of those mysterious harmonies which Gauss admired in the properties of whole numbers. This was the creation by E. E. Kummer (1810–1893), Dedekind, and Kronecker of the theory of algebraic numbers. Is this particular field Kummer's invention of ideal numbers is comparable to that of non-Euclidean geometry, and likewise for Dedekind's creation of the theory of algebraic numbers which, as we noted [5.6], was the source of several ideas in modern abstract algebra. Another striking advance was the brilliant development of the analytic theory of numbers since the early 1900s. Of isolated problems inherited from the past that have been successfully grappled with, we may mention in particular E. Waring's (1734–1798) of the eighteenth century, to which we shall recur, and so for C. Goldbach's (1690–1764) conjecture. Another result of singular interest was the proof that certain numbers are transcendental, and the construction of many such numbers. We shall briefly indicate the nature of all these things presently. These preliminaries may well be closed with the following quotations and a note on amateurs of arithmetic.

The higher arithmetic [Gauss wrote in 1849] presents us with an inexhaustible storehouse of interesting truths—of truths, too, which are not isolated, but stand in the closest relation to one another, and between which, with each successive advance of the science, we continually discover new and wholly unexpected points of contact. A great part of the theories of arithmetic derive an additional charm from the peculiarity that we easily arrive by induction at important propositions, which have the stamp of simplicity upon them, but the demonstration of which lies so deep as not to be discovered until after many fruitless efforts; and even then it is obtained by some tedious and artificial process, while the simpler methods of proof long remain hidden from us.

In the preface to his *Introduction to the theory of numbers* (1929), L. E. Dickson (1874–) says,

During twenty centuries the theory of numbers has been a favorite subject of research by leading mathematicians and thousands of amateurs. Recent investigations compare favorably with the older ones. Future discoveries will far surpass those of the past.

Among other things I have included from the vast mass of results accumulated during the past 2,500 years are several items that still intrigue amateurs. The simplicity of their statements is no index of their difficulty. In fact it seems almost as if amateurs have been attracted by some of the hardest problems that continue to rebuff professionals. Occasionally some observant amateur notices an interesting property of numbers that the professionals had overlooked. The professionals then have new problems to worry them, sometimes for a century or more.

❖ 11.2 ❖ *Fermat and Mersenne Numbers*

The theory of numbers as an independent discipline of mathematics originated with P. S. (de) Fermat (1601–1665). By profession Fermat was a jurist and parliamentarian. His diversion was mathematics, which he approached as an amateur but developed as a master of masters. Although he was one of the founders of analytic geometry and the calculus [Chapter 15] he is remembered today principally for his work in the theory of numbers. He was the born arithmetician, and it is doubtful whether his penetrating insight into the intrinsic properties of the natural numbers 1, 2, 3, . . . has yet been equalled; certainly it has not been surpassed. It should be remembered that when Fermat did his greatest work he had only a clumsy substitute for the concise and creative algebraic symbolism which today is learned in a few weeks by beginners in school. Some of his deepest discoveries were reasoned out verbally with very few if any symbols, and those for the most part mere abbreviations of words. Any impatient student of mathematics or science or engineering who is irked by having algebraic symbolism thrust on him should try to get on without it for a week.

Fermat, Pascal, and other French mathematicians of the early seventeenth century had a mutual friend in Father M. Mersenne (1588–1648), who acted as a sort of post office, transmitting the mathematical and scientific letters of his friends from one to another. Mersenne also was an amateur. He survives in arithmetic through the numbers named after him. Some historians have conjectured that it was Fermat who really was responsible for these famous and mysterious numbers. But this seems unlikely, if for no other reason than that Mersenne was not given to stealing his friends' ideas.

Fermat's numbers, denoted by F_n, are defined by

$$F_n = 2^{2^n} + 1, n = 1, 2, 3, \ldots ,$$

so that F_4, for instance, is $2^{16} + 1$, F_6 is $2^{64} + 1$; thus

$$F_1 = 5, F_2 = 17, F_3 = 257, F_4 = 65,537,$$

all four of which are readily verified to be primes. The next is the sizable number

$$F_5 = 4,294,967,297.$$

Fermat's reason for investigating these numbers was his hope of finding a formula involving the integer n which for

$n = 1, 2, 3, \ldots$ would yield *only* prime numbers. He mistakenly thought that F_n was such a formula, but said explicitly that he could not prove that F_n is prime for all n. This is of crucial historical importance, because Fermat stated many of his results without saying how he had derived them and without indicating proofs, although he either said or implied that he had proofs. With one very famous exception, to be noted later, all of his asserted theorems have been proved. He was mistaken in his conjecture about F_n, and this one definite misjudgment is presumptive evidence that he knew when he had a proof and when he had not.

Euler (1732) factored F_5,

$$F_5 = 641 \times 6,700,417.$$

Next (1880), F_6, which is too long to write out here in unfactored form, was shown to be not prime,

$$F_6 = 274,177 \times 67,280,421,310,721.$$

Later it was proved that F_n is not prime for

$$n = 7, 8, 9, 11, 12, 18, 23, 36, 38, 73.$$

With a few reasonable assumptions regarding typography and a little common arithmetic, supplemented by logarithms if convenient, the reader may convince himself that if F_{73} were written out in full not all the libraries in the world could hold it. The last prime F_n so far found is F_4, and some hardy guessers have conjectured that there are no more primes F_n. The problem of proving or disproving that there are only a *finite* number of primes F_n is open. If there are infinitely many primes F_n this, conceivably, might be proved by some simple device such as Euclid's in his proof [1.3] that there is no end to the primes 2, 3, 5, 7, 11, 13, ... In this same direction F. M. G. Eisenstein (1823–

1852), a first-rate arithmetician, stated (1844) as a problem that there are an infinity of primes in the sequence

$$2^2 + 1, \; 2^{2^2} + 1, \; 2^{\overline{2}^2} + 1, \; \ldots \ldots$$

Doubtless he had a proof. This looks like the sort of thing an ingenious amateur might settle. If anyone asks why I have not done it myself—I am neither an amateur nor ingenious. In passing, I recall the extraordinary statement attributed to Gauss, "There have been only three epoch-making mathematicians in history, Archimedes, Newton, and Eisenstein."

The Fermat numbers offer a beautiful example of "the new and wholly unexpected points of contact" (between the theory of numbers and other departments of mathematics) remarked by Gauss in his tribute [11.1] to arithmetic. He may have been thinking of the spectacular discovery which he made at the age of eighteen and which decided him to devote his genius to mathematics rather than to languages and philology—for which he had a rare gift. A geometrical construction using only a straightedge and a pair of compasses is called *Euclidean*. A polygon of which all the sides are equal and all the angles are equal is said to be *regular*. By 350 B.C. the Greeks knew Euclidean constructions for the regular polygons of 4, 8, 16, . . . sides and for those of 3 and 5 sides—the equilateral triangle and the regular pentagon. From these it was easy to construct regular polygons of, $2^c \times 3$, $2^c \times 5$, $2^c \times 3 \times 5$ sides, where c is any positive integer, and the Greeks in effect showed how. They got no farther. Young Gauss proved that if N is of the form 2^c or 2^c times a product of *different Fermat primes* F_n, then there is a Euclidean construction for a regular polygon of N sides. This form of N is both necessary and sufficient for the possibility of a Euclidean construction. It would be interesting if F_4 really is the last Fermat prime

—it is always satisfying to know that some problem is definitely done with. Simple Euclidean constructions for the regular polygons of 17 and 257 sides are available, and an industrious algebraist expended the better part of his years and a mass of paper in attempting to construct the F_4 regular polygon of 65,537 sides. The unfinished outcome of all this grueling labor was piously deposited in the library of a German university. Could misguided zeal go farther? It often has.

There is a baseless legend that Gauss's tombstone is inscribed with the diagram of his construction for the regular polygon of 17 sides. It may possibly be true that at one time Gauss, remembering the tomb of Archimedes with its diagram of the quadrature of the sphere, described by Cicero, wished such a memorial. If he had requested the like for the regular polygon of 65,537 sides, his executors might have had to duplicate or surpass the Great Pyramid. Of course this is an exaggeration. But nobody yet has been so pertinaciously stupid as actually to carry out the straightedge-and-compass construction for 65,537. Happily a Euclidean construction for 4,294,967,297 is impossible.

Mersenne's numbers M_p are as famous as Fermat's. They are defined by

$$M_p = 2^p - 1, \ p = 2, 3, 5, 7, 11, 13, \ . \ . \ . \ , 257, \ . \ . \ . \ ,$$

where p is prime. Mersenne asserted (1644) that the only *p*'s for which M_p *is prime* are

$$p = 2, 3, 5, 7, 13, 17, 19, 31, 67, 127, 257.$$

(If p were composite, $2^p - 1$ would be immediately factorable. For example, $2^6 - 1 = (2^3 + 1)(2^3 - 1)$, and so on.) There are 44 other primes p less than 257. So, according to Mersenne, M_p is not prime for any of these. It would be of great interest to know what grounds Mersenne thought he had for his assertions. It seems unlikely that he was just crudely guessing. He was an honest man and he was not a fool. Nevertheless he was far wrong. His first error was detected in the 1880s, when M_{61} was proved to be prime. This was shrugged off by those who still believed that the mysterious Father Mersenne really knew what he was talking about; 61 was just some careless copyist's mistake for 67. But in 1903 F. N. Cole (1861–1927) proved that M_{67} is not prime.

I should like here to preserve a small bit of history before all the American mathematicians of the first half of the twentieth century are gone. When I asked Cole in 1911 how long it had taken him to crack M_{67}, he said "three years of Sundays." But this, though interesting, is not the history. At the October, 1903, meeting in New York of the American Mathematical Society, Cole had a paper on the program with the modest title *On the factorization of large numbers.* When the chairman called on him for his paper, Cole—who was always a man of very few words—walked to the board and, saying nothing, proceeded to chalk up the arithmetic for raising 2 to the sixty-seventh power. Then he carefully subtracted 1. Without a word he moved over to a clear space on the board and multiplied out, by longhand,

$$193{,}707{,}721 \times 761{,}838{,}257{,}287.$$

The two calculations agreed. Mersenne's conjecture—if such it was—vanished into the limbo of mathematical mythology. For the first and only time on record, an audience of the American Mathematical Society vigorously applauded the author of a paper delivered before it. Cole took his seat without having uttered a word. Nobody asked him a question.

Another critical prime p was 257, the largest claimed by

Mersenne for his prime M_p's. D. H. Lehmer (son of D. N. Lehmer, whose factor tables for the first ten million numbers set an unsurpassed record for completeness and accuracy) proved (1931) that M_{257} is not prime, although the method used did not produce a pair of factors. In the fall of 1932 I was one of some friends of D. H. Lehmer who witnessed (in Pasadena) a test of his electronic machine for seeking the factors of large numbers. It was a dramatic demonstration of brains plus machinery, with the emphasis on brains. For a graphic account of that seance, the reader may consult D. N. Lehmer's Hunting Big Game in the Theory of Numbers (*Scripta Mathematica*, Vol. 1, pp. 229–235, 1932). As this is written, there are rumors that an electronic machine in Manchester (England) is being used to test Mersenne numbers.[1]

As someone may wish to push out farther, I state E. Lucas's (1842–1891) criterion, as much sharpened by D. H. Lehmer, for the primality of M_p. In the sequence

$$S_1 = 4, S_2 = 14, S_3 = 194, \ldots, S_{i+1}^2 = S_i^2 - 2, \ldots,$$

each term after the first is equal to the square of the preceding term minus 2. Thus

$$S_2 = 4^2 - 2 = 14, S_3 = 14^2 - 2 = 194, S_4 = 194^2 - 2$$
$$= 27634, \ldots,$$

and so on. The terms of the sequence increase with terrific rapidity, as the reader may see by calculating a few more. The criterion is: If p is a prime greater than 2, the Mersenne number $M_p = 2^p - 1$ is prime if, and only if, M_p divides the term of rank $(p - 1)$ in the sequence. Anyone who attempts to apply this to M_{257} will soon admit that some mechanical aid is desirable. Of course the terms of the

[1] (Added in proof.) Since the above was written, the machine has shoved the Mersenne conjecture out to primes p a little over 400. No further Mersenne prime was found.

sequence are not actually calculated out—even the hugest calculating machine in existence could not produce the sequence required for M_{257}. Short cuts provided by the theory of numbers make the criterion usable.

Having mentioned Lucas, I may recall that he was another of the great amateurs, in the sense that, although he was conversant with much of the higher mathematics of his day, he refrained from working in the fashionable things of his time in order to give his instinct for arithmetic free play. His *Théorie des nombres*, première partie, 1891 (all issued, unfortunately), is a fascinating book for amateurs and the less academic professionals in the theory of numbers. His widely scattered writings should be collected, and his unpublished manuscripts sifted and edited.[1]

When we come presently to Fermat's baffling 'Last Theorem' we shall note a curious connection between it and the Fermat and Mersenne numbers. For the moment I close Mersenne's account with immortality by quoting from D. H. Lehmer's paper in *The Bulletin of the American Mathematical Society*, Vol. 53, p. 167, 1947. M_p as before is $2^p - 1$, where p is prime.

(1) For the following primes p, M_p is prime,

2, 3, 5, 7, 13, 17, 19, 31, 61, 89, 107, 127.

(2) For the following primes p, M_p is not prime and has been completely factored.

11, 23, 29, 37, 41, 43, 47, 53, 59, 67, 71, 73, 79, 113.

(3) For the following primes p, M_p is not prime and two or more factors are known,

151, 163, 173, 179, 181, 223, 233, 239, 251.

[1] Some years ago the fantastic price of thirty thousand dollars was being asked for Lucas's manuscripts. In all his life Lucas never had that much money.

(4) For the following primes p, M_p is not prime and only one prime factor is known,

$$83, 97, 131, 167, 191, 197, 211, 229.$$

(5) For the following primes p, M_p is not prime but no factor is known,

$$101, 103, 109, 137, 139, 149, 157, 199, 241, 257.$$

In Lehmer's summary (6) the primes $p = 193$, $p = 227$ were undecided. H. S. Uhler (1872–), in 1948 settled these two. Neither is prime. Mersenne's final score, then, is five mistakes—the *inclusion* of 67, 257 and the *exclusion* of 61, 89, 107. What can the man have thought he was thinking about? It took 304 years to set him right.

From (3), (4), (5), it appears that something still remains to be done about Mersenne's mishaps. Modern calculating machines may do it—with the humble assistance of human brains.

❖ 11.3 ❖ *A Little about Primes*

The (rational) primes 2, 3, 5, 7, 11, 13, 17, 19, 23, . . . are the building blocks for the *multiplicative* division of the (elementary) theory of numbers. This is concerned with the consequences of *the fundamental theorem of arithmetic: A number* (positive integer) *is a product of primes in essentially one way only.* 'Essentially' means that two products of primes differing only in the arrangement of their factors are not counted as distinct. For example,

$$105 = 3 \times 5 \times 7 = 5 \times 7 \times 3,$$

and so on. The theorem can be proved by mathematical induction plus some simple but skilled ingenuity. E. Zermelo (1871–), whom we shall meet again toward the end of our whole story, gave such a proof in 1912 and published it in 1934. Others imitated him.

Of older theorems on primes, one of Fermat's may be specially mentioned, first because in itself it is prized as a gem by arithmeticians, and second because Fermat proved it by his method of *'infinite descent.'*[1] Every prime of the form $4n + 1$ is a sum of two squares in one way only. For example, $5 = 1 + 4, 13 = 4 + 9, 17 = 1 + 16, 29 = 4 + 25$, $101 = 1 + 100$. Moreover, as is seen immediately, no number of the form $4n + 3$ is a sum of two squares. The proof by descent assumes[2] that there is a smaller prime of the same form for which the theorem is false. Descending thus we reach the conclusion that the theorem is false for 5. But $5 = 1 + 4 = 1^2 + 2^2$. This contradiction establishes the theorem. The method of descent works infallibly when it can be applied, but usually, it is difficult to find the essential step down.

A cornerstone of the theory of numbers is Fermat's ('little,' or 'lesser') theorem of 1640, which states that if n is any integer not divisible by the prime p, then $n^{p-1} - 1$ is divisible by p. This (like nearly everything else inherited from the past in mathematics) has been extended and generalized in several ways, by no means all trivial. As simple proofs of the theorem are readily accessible in college algebras and elsewhere, I shall not describe them. Any student of modern algebra will know the far-reaching consequences of this theorem.

We remarked Fermat's attempt to find a formula yielding only primes. A closely similar problem is to find a criterion for primality. This was achieved before 1770 by J. Wilson (1741–1793), who stated that n is prime if, and only if,

$$1 + 2 \times 3 \times 4 \times \cdots \times (n - 1)$$

[1] Actually first used by Euclid.   [2] For lines omitted see page 437.

is divisible by n. For $n = 6$ this gives $1 + 120$, $= 121$, not divisible by 6, so 6 is not prime; for $n = 7$, $1 + 720 = 721$, divisible by 7, so 7 is prime. Unfortunately Wilson's absolute criterion is unusable for even very moderately large numbers—try it for $n = 101$.

There is an interesting connection between Mersenne primes and the so-called perfect numbers already mentioned [11.1]. Perfect numbers entered arithmetic with the Pythagoreans, who attributed mystical and slightly nonsensical virtues to them. If $S(n)$ denotes the sum of *all* the divisors of n, including 1 and n itself, n is called *perfect* if $S(n) = 2n$. This evidently agrees with the former definition [11.1]. For example, 6 is perfect since 1, 2, 3, 6 are all the divisors of 6, and the sum of these is 12, or 2×6. So for 28: all the divisors of 28 are 1, 2, 4, 7, 14, 28, whose sum is 56 or 2×28. Euclid and Euler between them proved that an *even* number is perfect if, and only if, it is of the form $2^c(2^{c+1} - 1)$, where $2^{c+1} - 1$ *is prime*. So to every Mersenne prime there corresponds an *even* perfect number. But what about *odd* perfect numbers? Are there any? The question was still unanswered in 1950 after about 2,300 years. Some progress had been made, however, in proving such things as that if an odd perfect number exists it must have at least six different prime factors as shown by Sylvester. Other negative results were proved in 1947 and later, but as they did not come near to settling the question I shall not describe them.

Almost the first question anyone, amateur or professional, might ask about primes is, "How many primes are there not exceeding any prescribed limit, say a billion, or x?" A more accessible question is, how many primes, *in the long run*, are there not exceeding x? More precisely, if $P(x)$

denotes the number of primes not exceeding x, is there a function of x, say $L(x)$, such that, as x tends to infinity—becomes indefinitely large—$P(x)/L(x)$ gets closer and closer to 1? If so, we say that $P(x)$ is *asymptotic* to $L(x)$. It was proved independently and almost simultaneously in 1896 by J. Hadamard (1865–) and C. J. de la Vallée Poussin (1866–) that $P(x)$ is asymptotic to $x/\log x$. (The log here is the natural logarithm.) This is the 'prime number theorem.' Its proof was a triumph of delicate mathematical analysis. Numerous modifications of the original proof engendered a vast literature of interest chiefly to specialists. Then, quite unexpectedly, A. Selberg in 1949 published an elementary proof. 'Elementary' does not mean 'easy.' The proof is elementary in a somewhat technical sense. It probably will be simplified.

Another famous theorem about primes is P. G. L. Dirichlet's (1805–1859) which states that in any arithmetic progression

$$an + b, \; n = 0, 1, 2, 3, \ldots ,$$

where a,b are positive integers having no common factor greater than 1, there are an infinity of primes. For example, there are an infinity of primes of the form $6n + 1$. Dirichlet proved this (1837) by difficult analysis. His proof was the real beginning of the modern analytic theory of numbers, in which *analysis*, the mathematics of *continuity* [4.2], is applied to problems concerning the *discrete* [4.2] domain of the integers. Again most unexpectedly, Selberg published (1949) an 'elementary' proof of this theorem of Dirichlet's.

On possibly a higher level of difficulty is the simple-looking question whether or not there are an infinity of primes of the form $n^2 + 1$. Almost anyone can guess about such problems that nobody yet has dented, much less split. Few reputable mathematicians today publish their unsubstantiated conjectures. It used to be imagined by some of the more romantic arithmeticians that the reckless guessers of the past possessed mysterious 'lost' methods of extraordinary power which enabled them to discover truths far beyond the reach of modern mathematics. The deflation of Mersenne [11.2] has made guessing unpopular in some quarters, but by no means in all.

A notorious guess was that of Goldbach in 1742. On only the scantiest numerical evidence he asserted that every even number greater than 2 is a sum of two primes, for example $30 = 13 + 17$. There was no significant progress toward a decision of this conjecture until 1937, when I. M. Vinogradov proved that every 'sufficiently large' odd number is a sum of three odd primes. Theoretically, the finite gap implied by 'large' could be closed in a reasonable time by modern calculating machines, and doubtless will be unless the exact Goldbach guess is disposed of before Vinogradov's theorem is outdated.

The prime-number theorem belongs to what is called the analytic theory of numbers. This vast and intricate structure, mostly a creation of the twentieth century, is largely concerned with determining the *order* (relative size) of the errors made if we take an *approximate* enumeration in a particular problem concerning a class of numbers instead of the exact enumeration. In this the leaders were Hardy, E. Landau (1877–1938), and J. E. Littlewood, (1885–). As a significant by-product of this analytic theory of (discrete) numbers, we may recall the extremely useful and simple concept of *order functions*, now a commonplace in treatises on analysis. This is indispensable in sufficiently accurate approximative calculation in many departments of applied mathematics, such as statistical mechanics, where *exact* results are sometimes humanly unattainable, even with calculating machines. The theory of *order* (magnitude of error, or of approximation) in this sense originated in 1892 in problems of the theory of numbers having no discernible connection with physics. Specifically, the 'O' function was introduced by P. Bachmann in his *Analytische Zahlentheorie* (1892). The physicist R. H. Fowler (1889–1943) told me that he acquired the great skill in approximative calculation exhibited in his *Statistical mechanics* of 1929 in a course at Cambridge under Hardy in the analytic theory of numbers.

The broader significance of all this work is its fusion of modern analysis and arithmetic into a powerful method of research in the theory of numbers.

❖ 11.4 ❖ *Diophantine Analysis*

As the name implies, this vast domain of the theory of numbers goes back to Diophantus of Alexandria, the earliest known master of the subject and, with Euler, Lagrange, and Gauss, one of the greatest. Diophantine analysis deals with the solution *in integers*, or *in rational numbers* (common fractions, [4.2]), of single equations or systems of equations in two or more unknowns.

The stated restrictions on the solutions are the sources of difficulty. For instance, in school algebra $2x + 3y = 5$ is solved for y thus, $y = (5 - 2x)/3$, where x may be *any* number. But *if only integers x,y are permitted*, the problem is not so easy. This particular equation can be solved 'by inspection': $x = 1$, $y = 1$ is a particular solution; if t is any integer, $x = 1 + 3t$, $y = 1 - 2t$ is the complete solution. If *positive* solutions are required, t must be zero. But, as the reader may satisfy himself, such an equation as

$$173x + 201y + 257z = 11,001$$

can hardly be solved by inspection.

There is a complete theory for systems of diophantine equations of the first degree. It is the final outcome of some fifteen centuries of effort by numerous arithmeticians beginning with the remote Hindus but essentially finished only in the 1860s by H. J. S. Smith (1826–1883). For equations of degree higher than the first, very little of any generality is known, and one of the outstanding unsolved problems of the theory of numbers—indeed of all pure mathematics, as emphasized in 1902 by Hilbert—is to devise usable criteria for distinguishing solvable from unsolvable diophantine equations. The corresponding problem for algebraic equations is at least approachable, if not yet solved in detail, by the Galois theory [9.8] and its modern extensions.

To keep this chapter within reasonable bounds I must omit much of interest and note only the most celebrated of all diophantine problems, Fermat's of about 1637. Fermat was accustomed to record some of his discoveries on the margin of his copy of C. G. Bachet's (1581–1638) edition of Diophantus' *Arithmetica*. What follows is Fermat's ever-memorable enunciation of his 'Last Theorem.' I quote Vera Sanford's translation of Fermat's Latin from *A source book in mathematics* (McGraw-Hill, 1929).

To divide a cube into two other cubes, a fourth power, or in general any power whatever into two powers of the same denomination above the second is impossible, and I have assuredly found an admirable proof of this, but the margin is too narrow to contain it.

Over three centuries of sustained efforts by some of the greatest mathematicians in history have failed to amplify that narrow margin to a complete proof that

$$x^n + y^n = z^n$$

is impossible in integers x,y,z all different from zero if n is an integer greater than 2. The exception $n = 2$ is necessary, as we noticed in connection with the ancient Babylonians [10.1]. Many a would-be disposer of the famous Last Theorem has wrecked himself by 'proving' the impossibility of the equation for all n's greater than 1.

In passing, may I request any reader of this section who imagines he has a proof not to send it to me? I have examined well over a hundred fallacious attempts, and I feel that I have done my share. One such, many years ago, stuck me for three weeks. I felt that there was a mistake, but couldn't find it. In desperation I turned the author's manuscript over to a very bright girl in my trigonometry class, who detected the blunder in half an hour. This was not as humiliating as it might have been. C. L. F. Lindemann (1852–1939), who in 1882 immortalized himself by proving the transcendence of π (to be noted presently), toward the end of his life published at his own expense a long alleged proof. The fatal mistake was almost at the beginning of the argument. Anyone contemplating a proof may be interested in what Hilbert said in 1920 when asked why he did not try: "Before beginning I should put in three years of intensive study, and I haven't that much time to squander on a probable failure."

Fermat left a proof by descent for the case $n = 4$, and Euler (1770) gave an incomplete proof for $n = 3$, subsequently completed by others. Since $x^{mn} = (x^m)^n$, and likewise for y^{mn}, z^{mn}, it would be sufficient now to prove the theorem for *odd prime* exponents n greater than 3. So far all attempts at a general proof have had to distinguish two cases: n does not divide any of x,y,z; n does divide one of them. The second appears to be much the more difficult. For the first case, J. B. Rosser (1907–) proved (1940) that the theorem is true for all odd primes not exceeding

41,000,000. This was bettered (1941) by D. H. Lehmer and E. Lehmer (1905–) to 253,747,889. For the second possibility, the limit up to 1950 was the 607 of H. S. Vandiver (1882–). For a summary of what was known up to 1946, the reader may consult the paper in the *American Mathematical Monthly*, vol. 53, 1946, pp. 555-578, by Vandiver, the arithmetician who since Kummer went more deeply than anyone else into Fermat's Last Theorem.

Three special results may be mentioned for their suggestive connections with problems that interested Fermat. A. Wieferich proved (1909) that, if in the first case there are solutions x,y,z, the prime n must be such that $2^{n-1} - 1$ is divisible by n^2. The only n less than 2,000 that fits is 1,093. Numerous similar criteria were discovered after 1909. From Wieferich's criterion, E. Gottschalk deduced (1938) that there are no solutions in the first case if n is a Fermat prime $2^{2a} + 1$ or a Mersenne prime $2^b - 1$. Is this historically suggestive? Or isn't it?

Finally, H. Kapferer (1888–) proved (1933) the astonishing result that the existence of a solution of the equation

$$z^3 - y^2 = 3^3 \cdot 2^{2n-2} \cdot x^{2n}$$

in rational integers x,y,z, any two of which have no common factor greater than 1, is equivalent to the existence of a solution of Fermat's equation

$$u^n - v^n = w^n.$$

❖ *11.5* ❖ *Algebraic Numbers*

The greatest service Fermat's Last Theorem so far has rendered mathematics was its instigation of the theory of algebraic numbers. This theory, as we have noted [5.6], was responsible for some of the guiding concepts—for instance, ideals—in modern algebra, and these in turn have reacted on modern mathematical physics.

The positive, zero, and negative whole numbers of common arithmetic are called *rational* integers, to distinguish them from *algebraic* integers, which are defined as follows.

Let a_0, a_1, a_2, . . . , a_{n-1}, a_n be $n + 1$ given rational integers, of which a_0 is not zero, and not all of which have a common divisor greater than 1. It is known from the fundamental theorem of algebra (first proved in 1799 by Gauss) that the equation

$$a_0x^n + a_1x^{n-1} + \cdots + a_{n-1}x + a_n = 0$$

has exactly n roots. That is, there are exactly n real or complex numbers, say x_1, x_2, . . . , x_n, such that if any one of these be put for x in the equation, the left-hand side becomes zero. Notice that no kind of number beyond the complex has to be created to solve the equation. If $n = 2$, we have the familiar fact that a quadratic equation has precisely two roots. For clarity I repeat that a_0, a_1, a_2, . . . , a_n in the present discussion are rational integers, and that a_0 is not zero. The n roots x_1, x_2, . . . , x_n are called *algebraic numbers*. If a_0 *is* 1, these algebraic numbers are called *algebraic integers*, which are a *generalization* of the rational integers, as seen in a moment. For instance, the two roots of $3x^2 + 5x + 7 = 0$ are algebraic numbers; the two roots of $x^2 + 5x + 7 = 0$ are algebraic integers.

A rational integer, say n, is also an algebraic integer, for it is the root of $x - n = 0$, and so satisfies the general definition. But an algebraic integer is not necessarily rational. For instance, neither of the roots of $x^2 + x + 5 = 0$ is a rational number [4.2], although both, according to the definition, are algebraic integers. In the study of algebraic numbers and integers we have another instance of the tendency to generalization which distinguishes modern mathematics.

Omitting technical details and refinements, I shall give some idea of a radical distinction between rational integers and those algebraic integers which are not rational. First we must see what a field of algebraic numbers is. At the risk of some repetition (see [5.7]) for clarity, I shall state the basic definitions.

If the left-hand side of the given equation

$$a_0x^n + a_1x^{n-1} \cdots + a_n = 0,$$

in which $a_0, a_1, \ldots, a_n$ are rational numbers [4.2], can *not* be split into two factors each of which has again rational numbers as coefficients, the equation is called *irreducible* of *degree n*.

Now consider all the expressions which can be made by starting with a particular root of an irreducible equation of degree n (as above) and operating on that root by addition, multiplication, subtraction, and division (division by zero excluded). Say the root chosen is r; as specimens of the results we get $r + r$, or $2r$, r/r or 1, $r \times r$ or r^2, then $2r^2$, and so on indefinitely. The set of all such expressions is evidently a field, according to our previous definition [3.1]; it is called the *algebraic number field* of degree n generated by r. It can be proved that any element of the field generated by r is expressible as a *polynomial* in r of degree not exceeding $n - 1$, where n is the degree of the irreducible equation having r as one of its roots. The field will contain algebraic numbers and algebraic integers. It is these integers at which we must look, after a slight digression on rational integers.

The *rational* primes are 2, 3, 5, 7, 11, 13, 17, 19, 23, 29, . . . , namely, the numbers greater than 1 which have only 1 and themselves as divisors. The *fundamental theorem of* (*rational*) *arithmetic,* I recall, states that a rational integer greater than 1 either is a prime or can be built up by multiplying primes in essentially *one way only* [11.3]. This is so well known that some writers of schoolbooks assert it to be 'self-evident,' which is a signal instance of the danger of the 'obvious' in mathematics. Whoever can prove it without cribbing from a book may have the stuff of a real mathematician in him. I have mentioned [11.3] Zermelo's proof.

Primes in algebraic numbers are defined exactly as in common arithmetic. But the 'self-evident' theorem that *every* integer in *every* algebraic number field can be built up in essentially one way only by multiplying primes is, unfortunately, *false.* The foundation has vanished and the whole superstructure has gone to smash.

One should not feel unduly humiliated at having jumped to this particular 'obvious' but wrong conclusion. More than one first-rank mathematician of the nineteenth century did the same. One of them was Cauchy, but he soon pulled himself up short. In *some* algebraic number fields an algebraic integer can be built up in more than one way by multiplying primes together. This is chaos, and the way back to order demanded high genius for its discovery.

The theory of algebraic numbers originated with Kummer's attempt to prove Fermat's Last Theorem [11.4]. About 1845 he thought he had succeeded. His friend Dirichlet pointed out the mistake. Kummer had assumed the truth of that 'obvious' but not always true theorem about the prime factors of algebraic integers. He set to work to restore order to the chaos in which arithmetic found itself, and in 1847 published his *restoration* of the fundamental law of arithmetic for the particular fields connected with Fermat's assertion. This achievement is usually rated as of greater mathematical importance than would be a proof of Fermat's theorem. To restore unique factorization into primes in his fields, Kummer created a totally new species of number, which he called *ideal*.

In 1871 Dedekind did the like by a simpler method which is applicable to the integers of *any* algebraic number field. Rational arithmetic was thereby truly generalized, for the *rational* integers are the algebraic integers in the field generated by 1 (according to our previous definitions).

Dedekind's 'ideals,' which replace numbers, stand out as one of the memorable landmarks of the nineteenth century. I can recall no instance in mathematics where such intense penetration was necessary to see the underlying regular pattern beneath the apparent complexity and chaos of the facts, and where the thing seen was of such shining simplicity. I refer here to what was said about ideals in connection with algebra [5.6]. A few evident verbal changes will give the definition of Dedekind ideals for an algebraic number field of any prescribed degree.

The first algebraic number field beyond the rational to be investigated was that generated by a root of $i^2 + 1 = 0$. The integers in this field are of the form $a + bi$, where a,b are rational integers. They are called *Gaussian integers*, as Gauss introduced them in 1828–32. The fundamental theorem of arithmetic holds for these integers, as also does the so-called *Euclidean algorithm* for finding the G.C.D. [5.5] of two numbers, but of course in a modified and generalized form. I state the kernel of the matter to prepare for a striking theorem proved in 1947. If we divide one positive rational integer by another, say 12 by 5, the least positive remainder, here 2, is less than the divisor 5. How could this be generalized to Gaussian integers? It is meaningless to speak of one complex number being less than another, because the complex numbers cannot be arranged in linear order. But if a unique *real* number [4.2] could be 'associated with' the complex number $a + bi$ [4.2], we might hope to continue.

As so often in the theory of numbers, it took unusual insight, not to say genius, to see the right thing to do. The *conjugate* of $a + bi$ is by definition $a - bi$; the product, $a^2 + b^2$, of these is called the *norm* of either of them. Gauss proved that in dividing one Gaussian integer by another the process can be so arranged that the norm of the remainder is always less than the norm of the divisor, exactly as in rational arithmetic. From this it can be shown that the fundamental theorem of arithmetic [11.3] holds for Gaussian integers. The field defined by $i^2 + 1 = 0$ is *quadratic*, since the defining equation is of the *second* degree, and *imaginary*, since the roots of the equation are not real numbers, and finally *Euclidean*, since, by the result just described, Euclid's algorithm for the G.C.D. holds for the integers of the field. How many quadratic fields are there in which there is a Euclidean algorithm? The question is really difficult. It was finally answered in 1947 by K. Inkeri, who proved that there are exactly twenty-two such fields. This is one of those completely satisfying results in the theory of numbers where a problem outstanding for many years is solved once and for all.

❖ *11.6* ❖ *Transcendental Numbers*

A mere glance at these must suffice. A number which is *not algebraic* is called *transcendental*. Otherwise stated, a transcendental number satisfies no algebraic equation whose coefficients are rational numbers. It was only in 1844 that the existence of transcendentals was proved, by Liouville. The transcendental numbers, hard as they are to find individually, are *infinitely more numerous* than algebraic numbers. This was first proved by G. Cantor, much to the astonishment of mathematicians.

A very famous transcendental is π (pi), the ratio of the circumference of a circle to its diameter. To 7 decimals $\pi = 3.1415926 \ldots$, and it was somewhat uselessly computed in 1874 to 707.[1] In 1882 Lindemann, whom we have already met [11.4] in a less favorable light, using a method devised in 1873 by Hermite, proved that π is transcendental, thus destroying for ever the last slim hope of those who would square the circle—although many of them don't know even yet that the ancient Hebrew value 3 of π was knocked from under them centuries ago.

In 1900 Hilbert emphasized what was then an outstanding problem, to prove or disprove that $2^{\sqrt{2}}$ is transcendental. The rapidity of modern progress can be judged from the fact that R. Kusmin in 1930 proved a whole infinity of numbers, one of which is Hilbert's, to be transcendental. The proof is relatively simple. Then, in 1934, A. Gelfond proved the vastly more inclusive theorem that a^b, where a is any algebraic number other than 0 or 1, and b is any irrational algebraic number, is transcendental. The proof of this is not easy.

❖ 11.7 ❖ *Waring's Conjecture*

Fermat proved that every rational integer is a sum of *four* rational integer squares (zero is included as a possibility). Thus $10 = 0^2 + 0^2 + 1^2 + 3^2$,

$$293 = 2^2 + 8^2 + 9^2 + 12^2,$$

and so on. In 1770 Waring guessed that every rational integer is the sum of a *fixed* number N of nth powers of rational integers, where n is any given positive integer and N depends only upon n. For $n = 3$, the required N is 9; for $n = 4$, it is known that N is not greater than 21.

---

[1] (Added in proof.) Since this was written, one of the new (ENIAC) calculating machines, in about 70 hours, computed π to 2035 decimals. Such a computation by hand might take all of a normal lifetime of hard labor.

Hilbert in 1909, by most ingenious reasoning, proved Waring's conjecture to be correct. But his proof did not indicate the number of nth powers required. Like much of Hilbert's work, the proof was one of 'existence.' This is slightly ironical, as the logic of existence theorems, among other things, was to force Hilbert in the 1930s to abandon his apparently promising attempt to prove the consistency of mathematical analysis. I shall recur to this in the concluding chapter.

In 1919, Hardy, applying the powerful machinery of modern analysis, gave a deeper proof, the spirit of which is appplicable to many other extremely difficult questions in arithmetic. This advance was highly significant for its joining of two widely separated fields of mathematics, *analysis*, which deals with the *uncountable*, or *continuous*, and *arithmetic*, which deals with the *countable*, or *discrete* [4.2].

Finally, beginning in 1923, Vinogradov brought some of these extremely difficult matters within the scope of comparatively elementary methods. Proceeding partly from Vinogradov's results, Dickson and S. S. Pillai almost simultaneously disposed of the entire Waring problem except for certain obstinate cases which were settled by I. M. Niven (1915–) in 1943. Conquests such as this would have seemed to the mathematicians of the 1840s to be centuries beyond them. I have dwelt on this episode—which after all is only an episode, though a brilliant one, in the general advance of mathematics—because it is a typical example of the internationalism of mathematics. A conjecture propounded by an Englishman was completely settled by the combined efforts of an Englishman, a German, a Russian, an Indian, a Texan, and a Canadian now a United States citizen. There is no national prejudice in mathematics.

Waring's problem belongs to the *additive* theory of numbers, in which it is required to express numbers of one specified class as *sums* of numbers in another specified class. For example, if each of the classes is that of all non-negative integers, we ask in how many ways is any given integer a sum of integers. This apparently idle question is less impractical than it may seem. It and its simpler variants are of use in statistical mechanics and the kinetic theory of gases. Euler initiated the theory of partitions in 1741. The modern theory falls into two main divisions, the algebraic and the analytic. The latter was elaborated in the 1920s and 1930s to facilitate actual numerical calculation. Here again asymptotic formulas were sought and found by advanced and difficult mathematical analysis.

❖ *11.8* ❖ *The Queen of Queens' Slaves*

Some of my unmathematical friends have incautiously urged me to include a note about the origin of modern calculating machines. This is the proper place to do so, as the Queen of queens has enslaved a few of these infernal things to do some of her more repulsive drudgery. What I shall say about these marvelous aids to the feeble human intelligence will be little indeed, for two reasons: I have always hated machinery, and the only machine I ever understood was a wheelbarrow, and that but imperfectly.

The first definitely recorded calculating machine (if we ignore our ten fingers and the abacus, which hardly rate) was Pascal's, completed in 1642—the year of Newton's birth, which is suggestive in view of what calculating machines may do for Newton's outstanding problem of three bodies [8.8] and its generalizations. The merit of Pascal's wheels and ratchets was that they performed the carrying of tens automatically. Further description should be superfluous for any normal youngster in our gadget-ridden age. Pascal, not living in a perpetual whirr of wheels within wheels, thought it necessary to guide his "dear reader" by the hand in leading him or her through the elementary mysteries of a train of gears. What cost the mathematical genius Pascal a tremendous effort is now an easily understood toy for any twelve-year-old with half a dollar to spend. Anyone with a quarter of a million to put out may buy an improved model and play with it at his leisure.

Pascal's gears were adapted to addition. Leibniz about 1671 invented a more versatile machine, capable, as he said, of "counting, addition, subtraction, multiplication, and division." But this was not Leibniz' main contribution to modern calculating machinery, which he made more or less in spite of himself and his bizarre theology. The main contribution was his recognition of the advantages of the binary scale, or notation, over the denary.

Probably it was not Leibniz, but may have been the ancient Chinese, who first observed that the simplest and inherently most natural way of representing integers is as sums of powers of 2, each power with a coefficient zero or one, instead of, as in our prevalent denary system, as a sum of powers of 10, each power with a zero or positive coefficient less than 10. In fact the Chinese either guessed or proved the special case of Fermat's 'lesser' theorem [11.3] that $2^{p-1} - 1$ is divisible by p when p is prime. This can be inferred from the representation of integers in the binary scale. But they slipped in supposing that this is a sufficient condition that p be prime. One of the first purely mathematical tasks of a certain modern machine devised for miliary purposes was an extensive correction of the ancient Chinese mistake. This of course was done while the machine was off duty. The machine did in a few hours what calculation 'by hand' would have taken several years.

To give an example of the two scales, 2,456 in the denary scale means $(2 \times 10^3) + (4 \times 10^2) + (5 \times 10) + 6$. In the binary scale no such sequence of digits as 2,456 can occur, since each digit in the number represented must be either 0 or 1. To express 2,456 in the binary scale we note the highest power of 2 that does not exceed 2,456. It is 2^{11}, and $2,456 = 2^{11} + 408$. We then proceed in the same way with the remainder 408, and so on: $408 = 2^8 + 152$; $152 = 2^7 + 24$; $24 = 2^4 + 2^3$; thus

$$2,456 = 2^{11} + 2^8 + 2^7 + 2^4 + 2^3 = 100,110,011,000$$

in binary notation.

There are easy routines for converting a number from any one scale to any other. In the 1890s some school algebras included a short chapter on scales of notation. It disappeared from the texts because it seemed to be useless. Today binary arithmetic is the mathematics behind one efficient type of 'digital' computing machines. These machines rely on simple counting. The binary scale may also be the ultimate secret of the creation of the universe. For the fact that zeros and ones suffice in binary arithmetic for the expression of any integer convinced Leibniz that God had created the universe (1) out of nothing (0). Leibniz was not only a great mathematician but a great philosopher as well. However, he confused 'nothing' with 'zero'—a remarkable feat for a mathematician and logician.

Between Pascal and Leibniz and our own times there was another pioneer who deserves mention, C. Babbage (1792–1871), who invented the first really modern calculating machine. Babbage was another of those British nonconformists, like De Morgan, who cared little for the fashions of his day. In his youth he was one of the brash young founders (1815) of the Analytical Society at Cambridge, where patriotic idolatry of Newton had put the British mathematicians a century behind their Continental rivals. The Society raised British mathematics from the dead and gave it at least a semblance of life until it almost succeeded in

committing suicide in the stupidest examination system—
except possibly the mandarin Chinese—in history. Babbage's 'analytical engine' was capable of tabulating the
values of any function and printing the results. Only a part
of the machine was ever actually constructed. The project
was financed by the British government because of possible
use by the Admiralty. The expense was far greater than had
been anticipated and the government refused to continue,
although a scientific commission reported favorably. I have
seen it stated somewhere that the government put £100,000
into the 'engine' before consigning it to the Kensington
Museum, but this is hard to believe. In spite of the commission's favorable report, it is doubtful whether the
machine tools of the time were adequate for finishing the
job with sufficient accuracy.

I shall not attempt to describe a digital computer, partly
for the reasons already given and partly because as this is
written new and improved types of machines are being
invented and manufactured in rapid succession. Commercial
competition has stimulated invention, as usual. The underlying mathematics is simple enough compared to the physical and engineering problems that must be solved for actual
production. Electronics is (was?) one of the most effective
sciences applied in design. I have already noted [5.2] the
connection with a two-valued logic, an open or closed circuit, a 'yes-no,' and hence a *binary* mechanism.

There is one feature in which some of the modern machines differ from all their immediate predecessors. They
exhibit a fair but far from perfect imitation of human
memory. Numbers are stored in the machine by one ingenious device or another for future use; the machine 'remembers' them and automatically puts them into the
calculation. The electronic tubes accept and retain numbers
for future use. But the electronic memory, like the human,

is finite, and a not completely solved problem (at this writing) is to improve the memory of a machine without diminishing its speed. Similar problems arise in the transmission of messages of any kind. There is one limitation of speed no machine is likely to overcome unless Einstein is wrong; the speed of light is an upper limit to the speed of whatever can move.

I shall only mention another type of machine, the 'analogue computers.' If there is a basic mathematics for these, it may be kinematics or topology. Older examples of analogue computers are slide rules, planimeters, differential analysers, and tide predictors. The analogue machines are not counters like the digital.

For those who wish further information on this topic there are two articles in the *Scientific American*, April and July, 1949—Mathematical Machines, by H. M. Davis, and The Mathematics of Communication, by W. Weaver (1894–). Interesting and impressive as these accounts are, I cannot see that the machines have dethroned the Queen. Mathematicians who would dispense entirely with brains possibly have no need of any.

❖❖❖❖❖❖❖❖❖❖❖❖❖❖❖❖❖❖❖❖❖❖❖❖❖❖❖❖❖❖❖❖❖❖

Chapter 12

ABSTRACTION AND PREDICTION

❖ 12.1 ❖ *From Maxwell to Radar*

It is time again for the Servant to get in a few words. As was remarked of Africa, there is always something new coming out of analysis. This vast domain comprises everything that concerns continuously varying quantities [4.2, 6.2]. Its importance for natural science is therefore evident, since it is true, apparently, that "all things flow." Fixity is an illusion, and analysis gives us a firm grasp on the laws of continuous change.

The progress in analysis since the time of Cauchy in the 1830s and well into the twentieth century was beyond all precedent. Today its scope is so vast that probably no mathematician is competent in more than a province or two of the entire domain. Particularly is this so if, as seems legitimate, we include under analysis the modern developments of differential geometry—the investigation of geometrical curves, surfaces, and so on, from the study of configurations and structure in a small neighborhood, as in the distance formulas for curved spaces already described [10.2, 10.3]. The last man to look out over the whole field of the analysis of his time was the universal-minded Poincaré, and he was able to do so largely because great tracts of analysis as it was in his day were his own creations. On practically every department of mathematics this outstanding genius left his deep impression.

In all of this bewildering progress it is not easy to find commanding points of view from which to survey any significant expanse of the whole unbounded territory. Temporary boundaries in all directions are being pushed forward so rapidly that the eye soon loses them in the distance, and this activity shows no sign of abating.

Nevertheless there have been one or two general directions of advance, at which we must look. It may be said that three of the leading activities since the 1870s and through the 1940s were the invention and exploitation of new species of functions in almost inconceivable variety, continual generalization, and drastic criticism of the foundations on which analysis rests.

Standards of rigor in proof after the 1860s were constantly raised. What had passed as satisfactory at an earlier period was minutely scrutinized, often found to be shaky, and firmly established according to the standards of the day. In this direction finality is no longer sought, for it is apparently unattainable. All that we can say is, in the words of E. H. Moore (1862–1932), "Sufficient unto the day is the rigor thereof."

Another tendency, which doubtless will persist, manifested itself all through the period from the late 1860s to the 1940s. No sooner was a significant advance made in another department of mathematics than analysis seized upon the central ideas and assimilated them with voracious speed. Thus groups, invariants, much of geometry, and parts of the higher arithmetic and modern abstract algebra successively become its more or less willing prey. On the other hand, wherever it was found possible to apply the techniques of analysis to any other domain, whether purely mathematical or scientific, the advance was swift and sure. We saw an example in the theory of numbers.

Nowhere more strongly than in analysis do we appreciate the peculiar power of mathematical reasoning. This power is traceable, at least partly, to the fact that mathematics does not direct isolated or individual weapons at a problem, but unites whole complexes of subtle and penetrating chains of thought into new, intimately wrought engines of reason, often expressed by a single symbol whose laws of operation are once for all investigated, and then applies these *as units* to the problem in hand. It is somewhat like the advance of an entire well-coordinated army obeying a single order; the commanding general does not fuss over the details by which the individual companies are to maneuver. The mere creation of the single weapon begets unsuspected power in the parts of which it is composed, and operating as a unit, the whole achieves incomparably more than the sum of the achievements of the parts. Unsuspected possibilities present themselves automatically. Before the designer of the new weapon is aware of it, he has made a conquest of which he never dreamed.

Instance after instance of this peculiar power might be cited. We have already seen an example in the verified predictions of general relativity. Another famous example will be given in the next chapter, and two more immediately. In each it was not mathematics alone that won the victory. The insight, or intuition, of a great scientist was in each case necessary before the physical problem could be formulated mathematically. But none of these advances could have been made—certainly none *was* made—without powerful mathematical analysis. The ability to translate scientific problems into mathematical symbols appears to be as rare as the genius which creates the mathematics to solve the problems.

My present example goes back to 1864. In that year Maxwell (1831–1879), having translated some of M. Faraday's (1791–1867) brilliant experimental discoveries in electromagnetism into a set of differential equations, and having filled out the set of equations to fit a physical hypothesis of his own, proceeded to manipulate the equations according to standard processes of mathematical analysis.

Now, one of the fundamental equations in mathematical physics expresses the fact that whatever satisfies the equation in a given instance is propagated throughout space in the form of waves. Moreover, the equation contains the velocity with which the waves are propagated.

Manipulating his electromagnetic equations, Maxwell derived from them the wave equation of mathematical physics. The indicated velocity was that of light. (The equations will be stated in [17.4].) Whether he was surprised at what the mathematics gave him, he does not record. At any rate he proceeded to exploit his discovery in grand fashion. He showed that electromagnetic disturbances must be propagated through space as waves. Further, from the manner in which the velocity entered the equation, he concluded that light is an electromagnetic disturbance. We shall come back to this in Chapter 17.

This was in 1864. Maxwell died in 1879. In 1888, H. Hertz (1857–1894), directly inspired by Maxwell's prediction of 'wireless' electromagnetic waves, and guided by his predecessor's mathematics, set out to produce the waves experimentally and to determine their velocity. From his success have sprung the whole wireless and radio and television industries of today, and they all go back to a few pages of mathematical analysis. But again we must emphasize that without Maxwell's extraordinary skill in *setting up the equations* and his physical intuition, the mathematics could not have got very far. On the other hand Hertz might never have even started.

The end of the story is not yet in sight.

The abstruse theories of Maxwell and the crude experiments of Hertz are still bearing fruit in dozens of unexpected ways [I quote

L. A. DuBridge, who during World War II was in charge of the Radiation Laboratory at the Massachusetts Institute of Technology, where Loran was invented and developed and where great advances in the development of radar were achieved] . . . Loran offers for the first time in the history of navigation the possibility of putting the stars out of business as the sole guide to the mariner. Already [1949] the principal sea lanes of the world are blanketed with a series of radio signals which a proper receiving set can convert almost instantly into a navigational 'fix'—day or night, in clear weather or foul.

Loran makes use of the simple fact that radio waves travel always with a fixed known speed. Hence if three suitably located synchronized transmitting stations simultaneously send out a sharp pulse signal, the intervals between arrival of these signals at a ship will depend on the location of the ship, that is on its relative distance from the three stations. Hence by timing the interval between the arrival of the first signal and the other two, the ship's location can be accurately determined . . . .

Let me turn briefly now to another application of radio waves to sea navigation—the technique known by the synthetic name 'radar.' While Loran is intended to supplant the stars in giving latitude and longitude, radar is designed to supplant the human eye in 'seeing' surrounding objects, such as nearby land masses, neighboring ships, uncharted rocks and even approaching storms.

Radar is a 'magic eye' indeed. Its principle of operation is simple. Radio pulses are sent out at short wavelength in a beam that scans a desired area. The echoes which are returned by objects in the beam are displayed on a screen, which provides an accurate map of the given area.

On board a ship the radar screen presents to the pilot a map of the nearby shore line, so that in darkness or in fog he can still navigate safely near the coast, or even into the harbor. Nearby ships or other obstacles are clearly displayed and the dangers of collision in darkness or fog are enormously reduced. Eventually radar beacons can be installed on the shore or over submerged rocks to identify more clearly and more distinctively the principal points serving in the role of lighthouses or channel lights, but being as visible in foggy weather as in clear. The days when a dense fog completely paralyzes shipping in busy harbors, or in locks and canals, may be passing.

DuBridge was talking of the applications of radio waves to marine navigation. But with a few evident amplifications his observations are good for aerial navigation. Guided missiles and hell in general are another application of Maxwellian waves. And it should never be forgotten that without radar, instead of "There will always be an England," it might well have been "There will always be a Germany." It was radar that discouraged the aerial bombing of Britain.

A personal reminiscence may be of interest here. Before Hitler had well started, the United States Navy sent a delegation of its scientific experts to tell the staff of a certain scientific institute what the Navy wanted the scientists to supply. The admiral heading the delegation described exactly the performance of what we now call radar. That was one of the things high up on his list of wants. He said it would be useful in the inevitable clash with Japan. Although what was to evolve into radar had already been invented in Britain, none of the men present seemed to have heard of it or would admit that they had. But some assured the admiral that they would give him what he wanted if he would give them time.

My second example is from chemistry—or atomic physics. It ranks with D. I. Mendelyeev's (1848–1907) triumph in his periodic law of the 1870s for the properties of the chemical elements. The modern quantum theory, largely an affair of pure mathematics as we have noted, predicted in 1927 that the familiar hydrogen of the chemical laboratories was not *an element*, but a mixture of *two elements*. The predicted two were sought and found experimentally— ortho-hydrogen and para-hydrogen.

With these two examples of mathematical prediction we may ask, as some philosophers do, what is the source of this prophetic power of mathematics when directed by a master?

For in spite of our mystification at its unforeseen triumphs, mathematics can tell us nothing that we did not, somehow, put into our assumptions from which all our ingenious mathematical tools are fashioned. How then does it appear to do so? I may recall in passing that the famous biologist T. H. Huxley (1825–1895), 'Darwin's bulldog,' somewhat rudely posed this question in the 1870s, and in so doing provoked Sylvester into one of his more florid testimonials for the glory of mathematics. Sylvester, in spite of his rhetoric, won the scholarly debate. Mathematics was vindicated, from twice-two to matrices, then an impractical frill on a supposedly useless algebra.

According to some, it is the scientific insight of the man who first constructs the equations to fit a particular physical situation that accounts for everything. Looking steadily at nature and seeing it whole, this childlike gazer knows intuitively what can be ignored without making his simplified mathematical picture of reality a grotesque caricature. Thus he takes the critical first step, that of idealizing his problem so that mathematics can rematerialize it for him in comprehensible imagery. As he has preserved all that is essential to his purpose, the unseduceable fidelity of mathematics will present him with equally just conclusions, which in their turn will truthfully portray aspects of reality which he sensed but did not consciously perceive.

Whatever be the explanation, it is historically true as we have just seen that scientists gifted with mathematical imagination have frequently made discoveries of unsuspected physical phenomena. To this extent at least mathematics has helped science. On the other side is the indisputable fact that the beautiful symmetry and simplicity of certain mathematical theories have caused them to be retained in science long after they should have been discarded to make way for increasing knowledge which they could not accommodate. A classic historical example will be noted later when we come in the next chapter to the rivalry between the Ptolemaic and Copernican descriptions of the solar system. But on the whole the credits appear to have greatly outbalanced the debits.

❖ *12.2* ❖ *Two Methods*

It is customary to call mathematics a science and, after the positivist philosopher I. A. M. F. X. Comte (1798–1857), to place it first in the classification of the sciences. So long as we remember the radical difference between mathematics and the physical or biologic sciences, no harm is done in calling mathematics a science. Something of the distinction between the mathematical method and the strictly scientific, however, must be seen before we attempt to uncover the mystery of mathematical prophecy in scientific discovery. The matter is extremely simple but none the less profound.

'If it is not abstract it is not mathematics' might be taken as a touchstone for discriminating between mathematics and other departments of precise investigation. A science has 'real' content, or claims to have. Electromagnetism, for example, is the organized body of information that has been acquired concerning electricity and magnetism as they 'actually' appear in human experience. Likewise astronomy, unless scrutinized too critically, is a systematized accumulation of 'facts' about the heavenly bodies, these celestial objects being assumed to have an 'existence' outside our 'sense perceptions.'

Until metaphysicians ask a physicist what value he attaches to such trite currency as 'matter,' 'motion,' 'the external world,' and 'fact,' he sees the universe about him clearly. Then his vision may begin to blur. Whatever doubts he may have, if he is an average man of science only mildly interested in the philosophy of his subject, will probably be swept aside, and he will get on with his work, confident that his science does have some content external to his own imaginings, even if he himself is unable to say precisely what that content is. For the majority of creative scientists are agreed that epistemological doubts, of great intrinsic interest philosophically, are not yet the main business of science.

With mathematics, however, it is entirely different. Few mathematicians with a modern training believe that their subject contains more than what they themselves put into it—as Huxley pointed out. So far as the external world is concerned, mathematics is as empty as a game imagined in a dream and forgotten on waking. Nothing whatever is in mathematics except the rules of the game, and these rules are prescribed at will by the player. This is merely the postulational method [2.2]. A mathematician may be blind, deaf, dumb, and paralyzed—but not brainless—so far as his experience of the external world is of consequence for his play.

The distinction between the mathematical method and the scientific is seen in the agonies of very young children to do what their teachers sometimes tell them is mathematics. Anyone who was subjected to elementary geometry when his infantile brain was as unripe as Verlaine's green walnut will recall the protracted misery he endured. Through stupid exercises of cutting out cardboard squares, rectangles, and circles, and measuring and weighing them, he struggled to placate his teacher by 'rediscovering' the idiotically simple 'rules' for finding the areas of such things. As scissor-and-balance gymnastics these tortures may have been an excellent initiation to the mysteries of a school laboratory in physics. As an introduction to mathematics, and in particular to geometry, they were silly, incompetent, immaterial, and irrelevant.

Experiment as scientists understand it is admittedly an indispensable adjunct to the progress of civilization and its probable destruction. But no eternity of such experiment will ever teach anyone what mathematics means. Sooner or later the cold plunge into pure abstraction must be taken if one is to learn to swim in mathematics and to reason as rational, thinking human beings do. Casualties among normal children occur only when overambitious pedagogues pitch their charges in too young, either drowning them outright or giving them a lifelong dread of all cold thinking. *The essence of mathematics is deductive reasoning from explicitly stated assumptions called postulates.* This is implied in what has been said [2.2] about the postulational method. To see its relevance for applied mathematics, we must look at it now more closely, to contrast it with what a nonmathematician might prefer.

Suppose that we knew nothing of geometry. Then we should be sensible to seek a usable set of rules for measuring our rectangular fields and city lots by experimenting with paper models. From a sufficiently large number of experiments we might conclude that, up to certain errors of observation, the area of a rectangle can be predicted, without cutting it up into unit squares, by multiplying the length by the breadth. This would give us a *statistical* 'law' for rectangles of precisely the same kind as those 'laws' widely applied in science, from nuclear physics to biology and sociology, where general principles are obscure or lacking.

All measurements in science are statistical in character. So far as any human being knows, an iron rod is never exactly 3 feet long in the laboratory. We say that the length is 3 feet with a 'probable error' of $\pm.00001$ foot, and we arrive at this conclusion by a long series of measurements carried out by physical devices, not by mathematical reasoning. In a similar manner, without understanding what 'intelligence' is, except that we can measure something to which we give the name 'intelligence,' we can predict with gratifying accuracy what percentage of a large number of human beings will score 80 per cent on a given set of questions. Our prediction is based on past scores in thousands of similar tests. Just as all iron rods manufactured by the same process are found on measurement to have the same length within narrow limits of error, so the tested group of human beings fits a fairly definite pattern, although we know nothing of them as individuals. Measurements such as these are the data to which mathematical reasoning is applied. A thorny metaphysics of measurement sprang up with relativity and the quantum theory, and it is possible that if we thoroughly understood the triangular relation between ourselves, measurements, and the things we measure, we might be able to dispense with the rest of metaphysics.

With the entrance of mathematics, the process just described is reversed. For brevity let us assume that long experience with fields and city lots has suggested the rule 'length times breadth' for the area of a rectangle. Instead of accepting this rule as an ultimate of experience we ask, as Thales did, whether it is a consequence of simpler rules. In this way we are led to formulate the postulates of elementary geometry. These postulates are so simple that they appealed to Euclid and others as 'self-evident' and 'true,' and in some intuitive sense as final. For example, the postulate that things which are equal to the same thing are equal to one another has been held to be 'self-evidently true.' Yet consistent—though rather poverty-stricken—systems can be constructed in which this postulate is either denied or ignored. Another postulate: no two straight lines can enclose a space. We saw this going wrong for the earth. Closer to the experience of surveyors is a third postulate of geometry: through any two fixed points one 'straight' line, and only one, can be drawn. This also goes wrong on the earth.

It may amuse the reader to recall what happened to the last two when a 'straight line' was defined as 'the shortest distance between two points.' On the surface of a sphere the 'straight lines,' according to this definition, are arcs of great circles, the geodesics [3.2] for the surface, when 'straight' means, as it should in common sense, 'geodesic.' Neither of the two 'self-evident truths' is true on a sphere. But, someone may object, straight lines are drawn on planes, not on spheres. What is wrong then with the 'shortest distance' or 'geodesic' definition? Nothing, except possibly that it was sublimated from experience at a time when intelligent human beings believed the earth was flat. Etymologically 'geometry' means 'earth measurement.' But 'geodesic' like geometry refers to the earth, etymologically, through the Greek for 'earth.'

The last would seem to substantiate the theory that all the score or so of postulates on which elementary geometry is founded are abstractions of common experience. On this theory geometry is a branch of physics. Other commonplaces of geometry and mechanics support the theory, for instance the concept of a 'rigid body.' Such a body can be moved freely about in space without alteration in any of its angles or lines. We noted that Euclid tacitly assumed this in his use of superposition. Although the 'common experience' theory accounts satisfactorily for the origin of geometry, it would not be accepted today as a valid description of the way in which the hundreds of postulate systems in modern mathematics have come into being. If 'experience' is to be credited here, it is experience at a higher level of abstraction—in brief, a questioning of abstractions prompted by the curiosity of mathematicians.

Having acquired its postulates, geometry parts company with physical experience and proceeds independently on a way of its own. Nothing further is injected into the subject beyond the familiar rules of deductive reasoning. These rules also have been ascribed by some to 'common experience,' but we need not go into this. We saw some of these rules in discussing symbolic logic [5.2], and that should be sufficient here.

Before many steps along the new road have been taken, bewildering accidents begin to happen. The 'self-evident' postulates begin exploding in bursts of beautiful or fantastic theorems which were by no means expected. Occasionally the spectacular pyrotechnics of the postulates may induce an uneasy feeling that somewhere along the road we have been deftly swindled into accepting dynamite for the staff of life. Possibly we have. Indeed the philosopher A. Schopenhauer (1788–1860) objected most peevishly to the Pythagorean theorem [10.1]. The proof, he declared, was as deceptive as a mousetrap into which he had been enticed by too easy assumptions. In his mathematics Schopenhauer was what is today called an intuitionist.

This feeling of getting something for nothing out of mathematics drives us back to physical experience. Suppose the truth of the Pythagorean theorem had never been suspected until geometers deduced it from their simple postulates. It is so amazing a geometrical deduction that its conformity to physical experience might well have been questioned by its discoverers. To reassure themselves they might have proceeded to dissect numerous beeswax models. They would then have made the mystifying discovery that the fruit of deductive reasoning is not recognizably different from that of physical experience. Having started with the simplest outline chart of everyday observations—as embodied in their postulates—the astonished geometers would have found that their first sketchy description included phenomena of which they had never dreamed. *Precisely the same sort of thing has happened repeatedly in science, from the Newtonian theory of gravitation to the quantum theory and relativity.* The so-called 'laws' of physical science—more properly, postulates—when subjected to mathematical reasoning have astonished scientists themselves with revelations of unknown worlds. We have just seen [12.1] an example in Maxwell's predictions.

The Pythagorean theorem of course had no such history —so far as is known. As a matter of fact nobody knows how it was discovered. It is usually thought to have been inferred by observation from special cases, in particular that when the sides including the right angle are equal, which 'leap to the eyes' in primitive designs of tile pavements. From these special cases the early geometers guessed the general truth of the theorem and then set out to prove it. They were thus forced to invent their system of geometrical postulates. Such is one historical theory.

The actual history of this particular theorem is unimportant for our purposes. The imaginary sketch does however portray the leading traits of scientific applications of mathematics.

First, observations of common or scientific experience are sublimated into abstractions. In this process only what is considered the most characteristic part of the observations is retained. Here judgment, taste, and opinion rule. Then to these abstractions mathematical reasoning is applied. Reason is dictator here. Next, the mathematical theorems resulting from this reasoning are deabstracted, or materialized, by a process inverse to that by which the observational data were abstracted in the first place. Human fallibility levies its heaviest tax at this stage. Finally the materialized theorems are compared with crude nature or with the results of experiment. Prejudice and the will to believe in ourselves usually collect here. If after this all-too-human ritual there is satisfactory agreement between what the mathematics predicted and what is found, the process is repeated on other phenomena of the same general order.

It is the incontrovertible verdict of scientific history that no such chain of abstraction, deduction, materialization, abstraction, and so on, has stood the strain of more than a hundred trials in the prediction of results that could be considered new. The majority have broken under far fewer. When a chain breaks it may be welded or thrown out on the scrap heap, the latter being the safer procedure. A classic instance is the theory of the all-pervading luminiferous ether, which after many modifications during the nineteenth century was finally discarded as a delusion early in the twentieth.

But not always is the broken chain discarded entirely. For some purposes, segments of it may still be as useful as they ever were. As a current instance, the Newtonian theory of gravitation is sufficient for nearly all the relatively unrefined observations to which it was applied from Newton's day to Einstein's. For such observations it is retained and used. Quantitatively, it gives a sufficiently close description of what is observed in a suitably restricted region of 'space-time' [10.3]. Qualitatively, Einstein's theory differs radically from Newton's and in this respect Newton's theory has been superseded. As another instance of a once-accepted description which has been discarded—this one without a conservative reservation as to its approximate correctness—we recall the phlogiston theory of heat. The two-fluid theory of electricity was likewise discarded because it failed to check with observation, and so for the elastic-solid theory of light—a structure of rare mathematical beauty. And so on. If anything more nearly permanent than the names of

these great—and, for their respective epochs, fertile—
mathematical systems of the past has survived as acceptable descriptions of the observable universe, it is but barely
discernible in current science. Not everything, however, has
gone. What has lasted is a great mass of useful mathematics
devised originally to develop theories now either modified
out of recognition or discarded as no longer adequate in a
continually progressing science.

A fairly recent historical example of the progress just outlined may be illuminating. After the motions of the planets
in our solar system had been worked out in detail more than
sufficient for commercial purposes—navigation, the construction of almanacs, tide tables, and so on—certain dynamical astronomers began a serious attack on the sidereal
universe. How were the inconceivably complex motions of
a swarm of millions upon millions of stars, such as the host
in our own galaxy (the Milky Way), to be unraveled? A
direct assault by Newtonian dynamics was out of the question. Even today there exists no usable mathematical
method for completely describing the motions of only three
bodies attracting one another according to the Newtonian
law of the inverse square of the distance. How then attack
millions?

It is a long story, and in the end the intended epic turns
out to have been a burlesque; so I shall indicate only the
general tactics of one engagement. For reasons that may
easily be imagined, the actual problem was replaced by a
highly idealized perfection of itself. Instead of regarding the
swarming stars as the gigantic bodies they probably are,
mathematicians perceived their host (in imagination only)
as a cloud of equal particles. In short, the galaxy was idealized into a perfect gas. The Gargantuan suns were the
molecules in this hypothetical vapor.

Now, a great deal was known about the observed and mathematical behavior of gases—the kinetic theory—when this application was made. Gases indeed appear to exist chiefly to give experts in statistical mechanics and the mathematical theory of probability an opportunity to appease their hunger for calculation. All this mass of kinetic mathematics was now laid down like a barrage on the idealized gaseous galaxy. Among other questions which the mathematics seemed heavy enough to reduce was this: "Is the galaxy a stable structure, or will it disperse in space, ceasing to exist as a coherent community?" The answer (I have forgotten what it was) is of no consequence. All we need observe is the chain of abstraction, deduction, materialization, and note next how the last of these links was forged.

The human race cannot hope to check the accuracy of its calculations on the Milky Way. The cosmic clock ticks in millions of years, and our kind will have vanished before the galaxy touches its noon. But ours is only one galaxy in hundreds of millions. The wrecks of others, dissipated like wisps of foam, may yet be visible in the unfathomable deep. These derelicts of eternity shall be our check.

Again it does not matter whether any of these cosmic wrecks were discovered. Since this grandiose attack miscarried, the cosmic dream has been altered beyond all recognition by the accession of new knowledge and more powerful methods of research, both mathematical and astronomical. The sketch of the process is, however, fairly typical. Beginning in observations, the scientific search passes through mathematics to new observations, and thus to the temporary verification or abolition of a scientific hypothesis. The pattern of all the miracles of prediction in science is the same.

❖ *12.3* ❖ *One Sort of Explanation*

It remains to suggest why the process of abstraction and generalization just described should work. Only a semanticist might be bold enough to doubt whether the question is meaningful, or is merely a senseless jumble of words like "Why is beauty hexagonal?" There seems to be no infallible rule for deciding whether a sentence in the correct grammatical form of a question is sense or nonsense.

Why should we expect to predict unknown phenomena? Many answers have been attempted. These will be found in the systems of the philosophers from Plato and Kant to Berkeley and Clifford, and from them, to mention only two from the twentieth century, to Whitehead and Russell. Explicit answers may not be evident, but the material for them is there. Perhaps the simplest and least satisfying of all explanations is Clifford's based upon 'mind stuff.' If everything from mud to metaphysics is a manifestation of mind, there should be no great difficulty in identifying mathematical symbols with matter. One objection which some raise to theories of this sort is that they make man out to be more conceited than he really is. Others object that all types of idealism must sooner or later drag in the 'Universal Mind,' as Berkeley did in his brand, in order that merely human seers shall be enabled to dream dreams and see visions at all. (Whitehead maintained that Berkeley is irrefutable.) These objectors—sometimes called materialists, as a term of opprobrium—assert that it is the duty of science to reject absolutely all theories immune to scientific investigation. According to these hardened skeptics, what the 'Universal Mind' is in desperate scientific need of today is a really close shave with W. Occam's (?–1349?) razor: "Entities shall not be multiplied beyond necessity."

I recently listened to, but did not understand, a lecture

by a distinguished philosopher in which six distinct theories to account for scientific prediction were expounded. The lecturer said there had been many more. He was not contradicted. There were professional philosophers in the audience. In the discussion following the lecture the sixty-three possible differences of opinion between the philosophers as to right or wrong were passionately debated. The discussion ended abruptly when a mathematical physicist objected that all were saying the same thing in different words. The philosophers were so deeply shocked by the possibility that any two of them, to say nothing of all, could agree on anything, that they subsided.

Fortunately for us, it is not necessary to take sides with either the materialists or the idealists. There is a simpler way out. Possibly this is no better than a cowardly evasion, but to avoid a scandalous embroilment with metaphysics I shall avail myself of the happy opportunity.

Looking out through the frosted windows of our senses, we may say, if we please, that we receive impressions of what we call the 'external universe' or 'reality.' But it is not necessary to say anything so mystical—at least for the present. It is sufficient to recall that certain scientific philosophers are content to start from the sense impressions themselves as 'reality,' without seeking for anything less familiar and more disputable. Einstein for one, in 1936, appeared to be satisfied to identify reality provisionally with the sense impressions which others refer to a yet more recondite 'reality.' Granting that the admitted difficulties blocking this approach are as massive as those obstructing any other, we can nevertheless see how abstract mathematics takes a grip on reality identified with the 'sense impressions' which we ourselves feel to be part of us.

An extreme example of this general type of theory was proposed in 1935 by Eddington, famous for his contributions to physics, astrophysics, and the philosophy of science.

Eddington's scientific faith startled some of his orthodox colleagues. "I believe," he said, "that all the laws of nature that are usually classed as fundamental can be foreseen wholly from epistemological considerations." The philosophy and physics suggested by this remarkable creed are elaborated in the posthumously published *Fundamental Theory* (1946). Anyone interested will find an account of this by E. T. Whittaker in the *Mathematical Gazette* (London), Vol. 29, pp. 137–144, 1945.

Accepting the identity of 'sense impressions' and 'reality,' we easily—perhaps too easily—account for all the rest. The first step is to schematize our sense impressions into pure abstractions which in some intuitive, simplified way we assume to follow a thought pattern similar to the structure of our sense impressions. From this schematization we proceed to seek a smaller number of assumptions from which the entire abstract thought pattern can be generated by deductive reasoning. These fewer assumptions may prove on examination to have a pattern or structure of their own. If so, the process is repeated, and the first assumptions in their turn are generated from a simpler set containing fewer assumptions. This process of successive abstraction we carry on as far as we can, the ideal being the deduction of everything we have constructed from a single grand assumption. This ideal, it need hardly be said, has not yet been attained in science, although its perfect functioning in mythology has been a commonplace for thousands of years.

At every stage of the passage toward the ideal we recede farther and farther from the evidence of our senses. In brief we ascend from sense to non-sense, and from there, according to the logical positivists, to nonsense. Concurrently we devise patterns that are progressively more amenable to mathematical reasoning. Having abstracted as far as we can, we reverse the process and compare the final result with reality—our sense impressions.

Chapter 13

FROM CYZICUS TO NEPTUNE

❖ 13.1 ❖ A Royal Road

Let us now return to a far past and look at one of the simplest sources of mathematicized science discovered by the ancient Greeks. They had no suspicion of the possible scientific importance of their discovery.

To one of Alexander the Great's (356–323 B.C.) tutors, the Greek mathematician Menaechmus, is attributed the discouraging remark, "There is no royal road to geometry." Alexander had impatiently ordered Menaechmus to abridge his proofs. Unable to oblige his impetuous pupil, Menaechmus nevertheless, perhaps in spite of himself, did succeed in leveling another royal road. This was the straight highway to the true beginning of mathematical astronomy and therefore also of analytical mechanics and mathematical physics. Without the purely mathematical inventions of this somewhat obscure Greek geometer, it is inconceivable that the course of the physical sciences, in particular mathematical physics and theoretical astronomy, could have followed even remotely any such direction as they actually have since the sixteenth century.

Of the life of Menaechmus little is known beyond his problematical dates, 375–325 B.C., and the uncertified tradition that he succeeded the incomparable Eudoxus (408–355 B.C.), precursor of the integral calculus, as director of the mathematical seminar at Cyzicus. Far more important for science than all the trivialities of Menaechmus' forgotten life is the memorable fact that he invented the conic sections. It was the simple geometry of these curves that led to the beginning of dynamical astronomy.

The conics are easily visualized. Imagine a cone standing on a circular base B. The surface of the cone is to be extended (as in the figure) indefinitely in both directions through the vertex V. The two parts of this extended cone issuing from V (one up, the other down), are called *nappes;* the straight line AVA' through V perpendicular to B is the *axis,* and any straight line, such as G, which passes through V and lies on the surface is a *generator* (see Figure 14). The curve of intersection of a plane with either or both nappes is a *conic section,* or briefly, a *conic.* According to this definition it is easily seen that there are precisely seven species of conics.

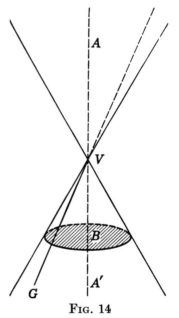

FIG. 14

In 1, 2, 3 the plane passes through V.

(1) If the plane passes through V and does not cut the surface elsewhere, the conic is a *point.*

(2) If the plane passes through V and touches the surface, the conic is a *straight line* (or pair of coincident straight lines).

(3) If the plane passes through V and also intersects B in distinct points, the conic is a *pair of intersecting straight lines* (see Figure 15).

In 4, 5, 6, 7 the plane does not pass through V.

(4) If the plane cuts the axis at right angles, the conic is a *circle* (See Figure 16).

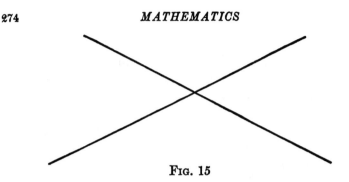

Fig. 15

(5) If the plane cuts all the generators in exactly one of the nappes, the conic is an *ellipse* (see Figure 17).

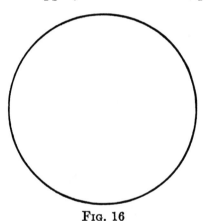

Fig. 16

(6) If the plane is parallel to a generator, the conic is a *parabola* (see Figure 18).

(7) If the plane cuts both of the nappes, the conic is a *hyperbola*, consisting of two branches (see Figure 19).

The first two of these are of no interest. Note, however, that the point conic can be considered as a circle with radius zero, and that the straight-line conic is a degenerate case of the pair of intersecting straight lines— when the lines coincide. For what is to follow, the ellipse is

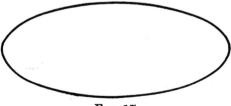

Fig. 17

the most interesting of the conics, although the parabola also has useful properties, two of which may be noted in passing.

A parabola is approximately the path of a ball, a bullet, or a rocket in the air. If the air offered no resistance, the path would be exactly a parabola. Thus if warfare were conducted in a vacuum, as it should be, the calculations of ballistics would be much simpler than they actually are, and it would cost considerably less than the hundreds of thousands of

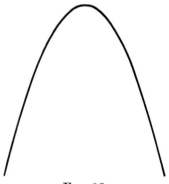

Fig. 18

dollars of taxes which it is now necessary to shoot away in order to slaughter one enemy patriot.

Parabolic mirrors offer a somewhat less bloody application of the conics, such mirrors being used in some automobile headlights. Suppose we were required to construct a mirror which would reflect the light from a point source in a beam of parallel rays. Trial and error might grope for centuries to discover what mathematics reveals with a turn of the hand: there is exactly one type of mirror which will do what is wanted, namely, the parabolic. Moreover the calculation

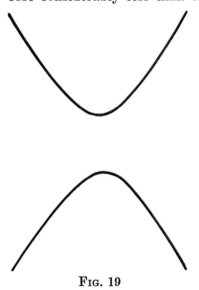

Fig. 19

prescribes the unique point at which the light must be
placed in front of the mirror to produce the parallel rays.
This point is called the *focus* of the parabola.

Passing to the ellipse, which will shortly assume the role
of guide in mathematical astronomy, we must define its
foci. First, to draw an ellipse, we tie a thread to two pins
stuck in the drawing board, say at F and F', and keep the
thread taut with the point P of the pencil; P then traces
out the curve—an ellipse—permitted by this restraint.

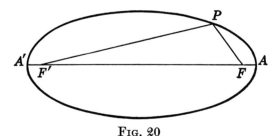

FIG. 20

Let the line joining F,F' cut the ellipse in the points A,A'.
Then the segment AA' is called the *major axis* of the ellipse,
and each of F,F' is a *focus* of the ellipse. From the manner
in which the curve was drawn it is clear that the sum of the
distances PF, PF' is *constant*—the same—for all positions
of the point P on the ellipse; this sum is equal to AA'. (See
also [7.4] and [10.1] for the sufficient data for finding the
Cartesian equation of the ellipse.)

If the foci F,F' move into coincidence, the ellipse degenerates to a circle. Thus, in a sense, an ellipse is a *generalization* of a circle. This is of some historical importance, as the
earlier Greek astronomers chose the too simple circle as
the key to the geometry of the heavens, whereas the ellipse
would have fitted more refined observations better.

It is also of interest in passing to record that at least some
of the early astronomers, following Plato (430–349 B.C.),

favored the circle on account of its supposed ideal perfection, not suspecting that a circle, with its smug rotundity,
is little better than a geometrical platitude. A circle no
doubt has a certain appealing simplicity at the first glance,
but one look at an ellipse should have convinced even the
most mystical of astronomers that the perfect simplicity of
the circle is akin to the vacant smile of complete idiocy.
Compared to what an ellipse can tell us, a circle has little
to say. Possibly our own search for cosmic simplicities in
the physical universe is of this same circular kind—a projection of our uncomplicated mentality on an infinitely intricate external world.

All the conics are unified from Descartes' algebraic point
of view of analytic geometry [7.2]. The several species of
conics correspond to the essentially distinct types of equations of the second degree in two variables.

❖ 13.2 ❖ *Kepler's Faith*

Through the labors of a host of Greek mathematicians,
of whom Apollonius (260?–200? b.c.) towers up as one of
the great geometers of all time, the geometry of the conic
sections was minutely elaborated long before the decline of
Greek learning. There is no evidence, as we have noted, that
any of the Greek mathematicians suspected that conics
would some day prove of paramount importance in the
dynamics of the solar system. The contrary appears to be
the case. Otherwise Ptolemy (second century a.d.) might
have been tempted to try ellipses instead of circles as a
clue to the geometry of planetary orbits. The Copernican
picture of the solar system with all the planets revolving
around the sun is sometimes said to have been imagined
(and forgotten) by the Greeks.

After the decline of Greece as a leader in geometrical research, knowledge of the conics was fostered and trans-

mitted to Christian Europeans by sagacious infidels, notably the Moslems. All through this protracted nightmare, the Ptolemaic description of the solar system, with the intolerable complexity of its cycles upon cycles, epicycles upon epicycles, brooded like a god. I need not describe Ptolemy's masterpiece. Except for its historical interest this massive work is now happily ignored. The Ptolemaic theory was one of humanity's major blunders in its gropings to touch what it cannot see, and also, it must be added, one of the major triumphs of creative mathematical thinking. If this seems paradoxical, we may recall from another field of imaginative thought Milton's (1608–1674) *Paradise Lost* and Dante's (1265–1321) *Divine Comedy*. Each in its time was esteemed as a revelation of divine truth. Though that time seems to have ended, both masterpieces last in their poetry. The enduring element of Ptolemy's system is its geometry.

Unfortunately for the progress of knowledge, the Ptolemaic theory, with the earth at the center of everything, was fully competent to account for the observed motions of the planets. This was its damnatory excellence. And more, with sufficient ingenuity the theory could be modified to accommodate certain new observations. Had not the simpler heliocentric picture of the solar system disclosed itself to Copernicus, we might still be admiring the Ptolemaic description of the solar system as one of the sublimest achievements of the human mind. Possibly it was; the Copernican revolution swept it into limbo.

Familiar as we are now with the rapid obsolescence of physical theories, it is difficult for us to imagine the terrific uproar occasioned by N. Copernicus' (1473–1543) fundamental remark that all the planets revolve round the sun. Nor perhaps did Copernicus himself foresee the full fury of the controversy his work was to precipitate when, in

1543, he touched on his deathbed the first printed copy of his treatise on the motions of the heavenly bodies. More prudent than his successor Galileo (1564–1642), Copernicus saved himself considerable embarrassment by resigning from life in the nick of time.

The bigoted intelligentsia of all colors, from humanists and theologians to astronomers and mathematicians, rallied to the flag of authority. Unlike the sophisticates of the twentieth century, the sages of the sixteenth seem to have preferred complexity to simplicity. They would have none of the beautifully direct heliocentric theory. The Chinese involution of the Ptolemaic system, which might have been buried the day Copernicus died, was embalmed and fitted with new gears to preserve it in a mechanical semblance of life for decades after it was dead. There was a similar brainless rush to the banner of tradition when Einstein replaced the Newtonian law of gravitation by another. Einstein escaped the concentration camps, it is true; yet—as we were assured by his former associates and compatriots—his theory is as trivial as was that of Copernicus.

After Copernicus the next long stride toward a rational celestial mechanics was taken by the astronomer Tycho Brahe (1546–1601). Either the laborious Tycho was too absorbed in his observations to have time for mathematics, or he lacked the right type of mind to synthesize his masses of data into a simple geometrical picture. What he had recorded of the motions of the planets, with an accuracy seldom approached before his time, ached for a mathematical interpreter. Did the planets revolve around the sun in circles, or was some less banal curve demanded to portray their orbits? J. Kepler (1571–1630), at one time Tycho's assistant, was to find the solution.

A highly gifted mathematician, a terrific worker, a first-rate astronomer, and a man of undeviating intellectual honesty, Kepler was the ideal candidate to sift the accumulated data, to calculate endlessly, and to be satisfied with nothing short of the completest accuracy attainable at the time. Undeterred by poverty, failure, domestic tragedy, and persecution, but sustained by his mystical belief in an attainable mathematical harmony and perfection of nature, Kepler persisted for fifteen years before finding the simple regularity he sought.

What Kepler accomplished is one of the most astounding feats of arithmetical divination in the history of science. His labors were rewarded by three of the grandest empirical discoveries ever made, *Kepler's laws:* all the planets move in ellipses round the sun, which is at one focus of these ellipses; the line joining the sun to any planet sweeps over equal areas in equal times; the squares of the periodic times of the planets are proportional to the cubes of the major axes of their orbits. All this he inferred by common arithmetic without benefit of logarithms.

The story of Kepler's epochal discoveries is so familiar that I need not dwell on it here, except to note the curious inconsequence of it all. What stimulated Kepler to keep slaving all those fifteen years? An utter absurdity. In addition to his faith in the mathematical perfectibility of astronomy, Kepler also believed wholeheartedly in astrology. This was nothing against him. For a scientist of Kepler's generation, astrology was as respectable scientifically and mathematically as the quantum theory or relativity or nuclear physics is to theoretical physicists today. Nonsense now, astrology was not nonsense in the sixteenth century. There have been attempts to show that Kepler did not really believe in astrology and practiced it only to help him make a living. The documentary evidence, however, apparently indicates the contrary.

What grounds have we for believing that any of the

spiritual stimulants which inspire men of science today to
believe in themselves and in their mathematical harmonies
of the universe are less absurd than the astrology and the
meaningless mysticism which sustained Kepler? If our own
stimulants are to be proved fraudulent in a century or two,
it does not follow that they are now worthless. Without
faith the spirit faints, and it makes not a particle of difference whether the necessary faith is generated by ecstatic
contemplation of a rag doll or by adherence to a prepos-
terous creed. The practical outcome is the same: a sublime
generalization concerning the cosmos—nothing less will
satisfy incorrigible seekers after finality—and a lugubrious
corollary bewailing the tragic destiny of man.

That such generalizations have frequently followed
Ptolemy's is of no importance for the immediate human
significance of scientific theories and in particular of the
imaginative creations of mathematical physics. Only when
we push mathematical mysticism to the limit and follow
Kepler into the realm of Eternal Verities do we let ourselves
and our successors in for a thoroughly deserved deflation.

Time after time the course is run over and over again in
the same even stride. A flash of intuition starts a strong
athlete fairly toward the winning post; surpassing his com-
petitors in the race, the man with endurance and faith in
himself wins; an accurate survey of the course shows that
the supposed winner never even distantly approached the
goal, and that a fresh start will have to be made. In the
meantime the treasury of knowledge is cluttered with bales
of unofficial records, and the mind of the devout believer in
everything from astrology to the latest transient speculation
of mathematicized physics is befuddled with superstitions
discarded long ago by reasonably alert spectators of the
race.

Again, we may ask whether Kepler would have discovered

his three laws if Tycho and others had been able to supply him with more accurate observational data. The answer is almost certainly "No." It is a fact that Kepler's discoveries fitted the available observations admirably. But it is also a fact that Kepler's calculations would not now check against observation with the precision demanded of modern astronomy. Nothing in the complicated heavens is simple enough to fit *exactly* the Keplerian arithmetic. Less honest investigators than a Kepler might easily persuade themselves (and the world) today that all the planets revolve round the sun in ellipses, and that the laws are indeed sustained by the more precise observations available since Kepler's time, but not Kepler. He would take account of the slight irregularities superimposed on his assumed perfect ellipses due to perturbations by all the planets, and would probably abandon the problem in despair on his deathbed. For it is true that no observed planet describes a perfect Keplerian ellipse as its orbit round the sun; the path is always slightly distorted. An elliptic orbit is correct only as a first very close approximation; it does not fit the observed facts exactly. What if Kepler could have used a modern calculating machine? I believe he never would have ventured to formulate his laws.

None of this is intended as a disparagement of Kepler's magnificent achievement. It is pointed out merely to indicate the way to another great triumph, and to emphasize that no set of scientific observations is as exact as its mathematical expression. Something is always discarded when the results of experiment are trimmed down to fit formulas and equations. That something, much or little, which is thrown away has frequently been of a scientific importance equal to what is retained in the mathematics. In some lucky instances the discarded part, when recovered and analyzed, has turned out to be of far greater scientific significance than what was kept. The situation is analogous to the re-working of the tailings from an abandoned gold mine when more economical processes or a rise in the value of gold make the venture profitable. So it was with what Kepler neglected and, after him, with the small discrepancy between the Newtonian theory of gravitation and the observed advance of the perihelion of Mercury [10.3].

The history of science counsels us to hold our mathematical generalizations regarding the universe lightly. But it also encourages us to anticipate that a first approximation may be valuable in suggesting a closer fidelity to the observables of nature. Kepler's magnificent approximations were such.

Before taking leave of Kepler, we may ask whether he would ever have formulated his laws had he not possessed the geometry of the conics elaborated by the Greeks. Here the answer is not so direct as in the case of more accurate observations. It is at least thinkable that a man of Kepler's patience would have been driven by his passion for accuracy to invent the ellipse had it not already been waiting for him. Indeed, Newton took a far more difficult step when he invented the calculus to enable him to analyze the subtleties of motion.

✤ 13.3 ✤ *Calculation plus Insight*

The full power of the mathematical method is not displayed in any such laborious cut-and-try as Kepler sweated over to find his three laws. The hidden strength of mathematics partly reveals itself at the next and far more difficult encounter with the unknown.

Before Kepler's laws could be deduced from something simpler, a mathematics of continuous [4.2] change had to be invented. This was the differential calculus (described in Chapter 15), brought to a usable state in the seventeenth century by Newton and Leibniz. In addition the empirical laws of motion had to be discovered and stated in a form adapted to mathematical reasoning. Galileo and Newton between them accomplished this. Finally, although it may not have been logically required, the concept of *force* had to be clarified, in particular the notion of a *force of attraction* between material bodies such as the sun and the planets. This was accomplished in Newton's law of universal gravitation. With these partly empirical preliminaries disposed of, the solar system was ready for idealization into a map of reality of sufficient abstractness to be explored mathematically.

The laws mentioned above may be recalled. First, the three laws of motion:

(1) *Every body will continue in its state of rest or of uniform motion in a straight line unless it is compelled to change that state by impressed force.*

(2) *Rate of change of motion is proportional to the impressed force, and takes place in the direction in which the force acts.*

(3) *Action and reaction are equal and opposite.*

The first of these defines *inertia;* the second (in which 'motion' means *momentum,* or 'mass times velocity,' both mass and velocity being measured in the appropriate units) introduces the intuitive notion of a *rate.* This last is probably the most important contribution of mathematics to astronomy and the physical sciences; explanation of what rates are must be deferred to Chapter 15. For the moment the common-sense notion of a rate suffices.

Newton's law of universal gravitation has a more audible mathematical ring:

(4) *Any two particles of matter in the universe attract one another with a force which is proportional to the product of the masses of the particles, and inversely proportional to the square of the distance between them.*

Thus, if m, M are numbers measuring the masses of the particles, and d measures the distance between the particles, the force of attraction is measured by

$$k \times \frac{m \times M}{d^2},$$

in which k is some constant number depending only on the units in terms of which mass, distance, and force are measured. If the distance is *doubled*, the attractive force is only *one-fourth* of what it was before; if the distance is *trebled*, the force is diminished to *one-ninth*, and so on.

It should appear reasonable even to one who remembers no mathematics beyond arithmetic that the path of one particle being attracted by another will depend upon the law of attraction between the two particles. If the attraction is according to the Newtonian law of the inverse square of the distance, the path will bear but little resemblance to that compelled by a law of the inverse third, or fourth, or fifth . . . power of the distance. It should also be fairly evident that the following question admits a precise answer: "If a celestial body, say a comet, is observed to trace a certain definitely known path, characterized geometrically, in its approach to the sun, what law of attraction between the sun and the body will account for the observed path?"

A similar problem was proposed in August, 1684, to Newton. For some months the astronomer E. Halley (1656–1742) and other friends of Newton had been discussing the problem in the following exact form: What is the path of a body attracted by a force directed toward a fixed point, the force varying in intensity as the inverse square of the distance? Newton answered instantly, "An ellipse." "How do you know?" he was asked. "Why, I have calculated it." Thus originated the imperishable *Principia* (*Philosophiae naturalis principia mathematica*, 1687) which Newton later wrote out for Halley. It contained a complete treatise on motion.

When solved fully by Newton, this problem answered both the direct and inverse forms of the question proposed, and accounted at one stroke for Kepler's three laws. The laws were deduced from the simple law of universal gravitation. The 'inverse' answer showed that if the path is an ellipse the law of attraction is the Newtonian.

It is definitely known that the synthetic, almost Greek, methods of the *Principia*, with their rigid Euclidean demonstrations, were not those by which Newton reached his results, but that he used his calculus as an immeasurably more penetrating instrument of exploration. The calculus [Chapters 14, 15] being a strange and prickly novelty at the time, Newton wisely recast his findings in the classical geometry familiar to his contemporaries. Today the deduction of Kepler's laws from the Newtonian law of gravitation is accomplished by means of the calculus in a page or two in the textbooks on dynamics. In Newton's day it was a task for a titan. We do seem to progress in some things, including scientific education.

Newton's law contained vastly more than Kepler's too smooth laws. Combined with the calculus, it made possible an attack on the observed irregularities of the supposedly perfect elliptical orbits. It is clear that if Newton's law is indeed universal, then every planet in the solar system must perturb every other and cause it to depart from the true ellipse which it would follow if it were the only planet in the system, and if both it and the sun were perfect homogeneous spheres. As the mass of the sun greatly exceeds that of the planets, the observed irregularities will be slight. But they will be none the less important. We shall see presently a spectacular example of the importance of attending to slight though awkward discrepancies between oversimplified mathematical perfection and obstinate facts of observation. Newton himself initiated the study of perturbations which was to enlarge our knowledge of the solar system and prove useful in the twentieth century in the yet more difficult field of atomic structure.

❖ *13.4* ❖ *Mathematical Prophecy Again*

The next episode in this brief story of the royal road opens with Newton in his forties and closes with a young man in his early twenties. As only one main road is being followed from the Cyzicus of Menaechmus to the Cambridge of J. C. Adams (1819–1892), we must pass by unexplored the many alluring highways branching off to other great empires of mathematical reasoning, and condense the boundless work of a century to one fleeting glimpse. At the end of the road we shall see a typical example—also one of the most famous—of the prophetic power of mathematics when applied to a masterly abstraction of nature. Newton's law of universal gravitation was such an abstraction; the methods of mathematical analysis [6.2] developed in the calculus elicited from Newton's law what it implicitly concealed.

Even today we look in vain for any generalization of physical science which has unified any such vast mass of diverse phenomena as was reduced to a coherent unity by Newton's law. At a first glance we may think that some of the relatively recent generalizations since the 1900s have a scope as wide as Newton's had in its heyday, but a little consideration shows that this is an illusion. Not only did universal gravitation as mathematicized in Newton's law sweep together, sift, and simplify the scattered astronomical knowledge of twenty centuries or more; for over two centuries it served as a suggestive guide in all fields of physical science where there was any glimmer of hope that a mechanical philosophy could simplify and unify the data of the senses.

Let us see what Newton's law (or postulate) of universal gravitation actually did. I transcribe the following summary from the *American Mathematical Monthly* (Vol. 49, pp. 561–562, 1942) from an article commemorating the tercentenary of Newton's birth.

From Kepler's first and second laws and his own three laws of motion, Newton deduced that each planet is attracted by a central force directed toward the sun, the intensity of the force varying inversely as the square of the distance between the two bodies; from his third law he deduced that all the planets are similarly attracted, the intensities depending on the sun's mass.

Almost as corollaries of these general conclusions, Newton showed how the sun's mass can be calculated in terms of the earth's mass: the length of any planet's year and its distance from the sun are the sufficient data. The mass of any planet having a satellite can be similarly computed. The same force of gravity that accounts for the fall of an apple was then shown to be sufficient for holding the moon in its orbit. Explicit definitions of 'mass,' 'gravity,' 'force,' and 'attraction' are unnecessary if the objective of these deductions is the correlation of known facts and prediction from them; the mathematical equivalents of the verbal statements suffice. Thus metaphysical disputes are short-circuited to their proper function.

The roll of the memorable conclusions Newton deduced from his law is only begun. The Newtonian theory of gravitation accounted for the tides. From the sun's mass, deduced from the theory, the height of the solar tide was calculated; as a sort of converse, from the observed heights of the spring and neap tides, the lunar tide was calculated, whence an estimate of the moon's mass was obtained. From

Newton's dynamics of a gravitating rotating body, it followed that the earth is not a sphere as had been supposed since ancient times, but an oblate spheroid, and the measure of flattening at its poles was calculated. Conversely, from the observed oblateness of any planet, the length of its day was shown to be calculable. Another easily verifiable prediction deduced from the polar flattening of the earth and centrifugal force was the variation of the weight of a body with the latitude. The attraction of the sun and moon on the earth's equatorial bulge, with similar attractions by the other planets, was proved to perturb the earth's axis of rotation by calculable amounts; and thus it was possible to follow "the wandering of the pole" and the precession of the equinoxes. In these and other problems, Newton inaugurated the theory and calculation of planetary perturbations.

The comets, which for ages had evoked the superstitious fear of savage and civilized man alike, were proved to be law-abiding members of the solar system, differing from the friendly planets chiefly in the smallness of their masses and the relatively high eccentricity of their orbits. (Their tails were explained only much later; their probable composition also was not determined until long after Newton was dead.) Further, it was possible to predict with high accuracy the dates at which some would return. Here was a deliverance from superstition that any lettered person could appreciate; and of all the deductions from Newton's hypothesis of universal gravitation, the accurately predicted return of a comet has been the most popular. Scientifically, however, it is not comparable with what followed from Newton's hard thinking on the problem of the moon's motion. Among its numerous claims to perpetual remembrance, this problem has the unique distinction of being the only one of which Newton confessed that it gave him a headache.

The motion of the moon presents a special case of the problem of three bodies. The moon's orbit is perturbed by the attractions of the earth and the sun, and also, to a lesser degree, by the attractions of the other planets. The perturbations cause irregularities in the moon's orbit, some of which had been observed by the Greeks and perhaps even by the Babylonians. None had been accounted for when Newton deduced them as consequences of universal gravitation, and in addition uncovered two more. In historical order from Hipparchus (second century B.C.) to Newton's contemporary J. Flamsteed (1646–1719), these were the equation of the center, the evection, the variation, and the annual equation; retrogression of the nodes, variation of the inclination; progression of the apses (only half the observed amount was given by Newton's calculation); and the inequalities of apogee and of nodes, Newton's discoveries. It is at this point that Newtonian gravitation makes one of its most direct contacts with the science of the twentieth century. The problem of three bodies remains an astonishingly prolific source of new mathematical methods and refined astronomical-physical observations.

The extreme simplicity of Newton's law itself is equaled only by one other powerful generalization of modern science, the law of the conservation of energy, which asserts that the total amount of energy in the universe remains constant. Electrical energy, for instance, may be transformed into heat and light, and seem to be dissipated, but actually nothing has been lost. This great generalization (now radically modified since the advent of relativity) may justly claim to have been a descendant of Newton's.

Some philosophers and historians of science would rank Darwin's theory of evolution with Newton's of universal gravitation, maintaining that Darwin's achievement was

as significant a step in the life sciences as Newton's was in
the physical sciences. Others assert that Newton's demanded a keener penetration than Darwin's for its successful accomplishment. Both have been modified with the
advance of science, but neither has lost any of its significance as one of the longest steps ever taken toward a rational understanding of the world in which we "move and
live and have our being." What Newton and Darwin did
affects all civilized men everywhere, as they are not affected by either their political or their religious creeds. A
Jew, a Christian, a Mohammedan, a Buddhist, an atheist,
a Republican, a Democrat, a fascist, a communist, a social-
ist, a pacifist, and a militarist can agree on Newton's law
in the vast domain for which it is still valid and useful, even
while killing one another to sustain their nonscientific
beliefs. Science makes no pretension to eternal truth or
absolute right; some of its rivals do. That science is in some
respects inhuman may be the secret of its success in allevi-
ating human misery and mitigating human stupidity.

As between Newton and Darwin, the theory of evolution
seems to have a deeper significance than the theory of universal gravitation in the beliefs even of those who never
heard of Darwin. We may forget that Newton's celestial
mechanics was once as repugnant to the traditionally
devout mind as Darwin's descent of man was not so long
ago. The controversies over celestial mechanics are forgotten; those over evolution are still remembered. Students
of science are sometimes disturbed by the so-called conflict
between science and religion. The neatest resolution of this
conflict I have yet heard was that of a Jesuit professor. He
had been lecturing on paleontology with its hundreds of
millions of years. A troubled student objected that the lecture disagreed with the account in Genesis. "Today is

Tuesday," the professor reminded him. "On Tuesdays we study science. The class in theology meets at this same hour on Thursdays." I shall continue as of Tuesday.

In the century following Newton a host of powerful mathematicians, including Euler, Lagrange, and Laplace, explored the heavens with Newton's universal law as their sole guide, finding almost everywhere measurably complete accord between theory and observation. The mathematical methods used by Newton's successors were not those of the *Principia*, but the more flexible analysis which evolved from the first rather crudely stated forms of the differential and integral calculus.

Not all these great mathematicians believed in the Newtonian law as a truly universal principle. Euler, for one, doubted whether anything so elementary could possibly account for even so simple a situation as that posed by the motion of the moon. By 'simple' here I mean simple only to uninstructed intuition. The motion of the moon offers one of the most complicated problems in the whole range of dynamical astronomy. It is a 'three-body problem' [12.3], the bodies being the sun, the earth, and the moon attracting one another according to Newton's law of gravitation. So great are the mathematical difficulties that any analyst of Euler's time might well have believed the Newtonian hypothesis of the inverse square law to be inadequate. The Newtonian law solves the problem of two attracting bodies completely and with a Keplerian simplicity. When more than two bodies attract one another according to the Newtonian law of universal gravitation, I repeat that no exact solution of the problem of completely describing their motions exists even today. The method of successive approximations is applied to yield progressively more accurate descriptions of the motion sufficient for all practical purposes, such as those demanded by the computation of the nautical almanac. As was remarked, modern calculating machines may indicate what are the facts for mathematics to take hold of, but no machine is likely to solve the general mathematical problem.

Among those who doubted the universality and adequacy of the Newtonian law was G. B. Airy (1801–1892), for long Astronomer Royal of Newton's own England. Airy's doubt was perfectly legitimate, and indeed highly creditable to a man of science, even if based on a total misconception of the nature of the real difficulties involved. But no skeptic has a right to impose the inertia of his disbelief on young men pushing forward to what they believe are attainable discoveries. To Airy more than to any other relic of academic conservatism is due the official indifference with which the calculations of young Adams—to be recounted in a moment—were received, when he ventured to sustain Newton by one of the most brilliant mathematical predictions in all the brilliant history of mathematical science. I have not space to go into the record of the official lethargy which robbed Adams of a unique 'first'; so with this brief (but on the whole adequate) obituary of his chief obstructor I shall pass on to what Adams did.

On March 13, 1781, W. Herschel (1738–1822) discovered with his telescope a new member of the sun's family, the planet subsequently known as Uranus after a happily abortive attempt to name it for King George III (1760–1820) of England. If the English conservatives had wanted to immortalize George, they should have called the planet America, or at least Boston. The newcomer offered a superb opportunity for testing the Newtonian law.

Before long, serious discrepancies between calculation and observation appeared in the orbit of Uranus. These could not be explained away by postulating faulty arithmetic or defective telescopes. Newton's hypothesis simply did not fit the observed facts.

But man is a theorizing animal, and the erratic irregularities of Uranus' motion were attributed to the perturbations of some more distant planet, then undiscovered. This hypothetical planet, attracting Uranus according to the Newtonian law of the inverse square, would account for everything, provided only that it existed and that its mass and orbit were of the right mathematical specifications to produce exactly those irregularities in the orbit of Uranus which had actually been observed.

Mathematically the problem was an inverse one of extreme difficulty. It would be laborious to calculate the effect of a *known* planet on the motion of Uranus, but the calculation would present no insuperable difficulty. But given the erratic motion it was much more difficult to reach out with mathematical analysis into the vague of ultraplanetary space and discover how massive, and where, the *unobserved* perturber was at any particular date.

By the early 1840s many astronomers believed that such an ultra-Uranian planet existed. In a historic prophecy Herschel stated on September 10, 1846, "We see it [the hypothetical planet] as Columbus saw America from the coast of Spain. Its movements have been felt, trembling along the far-reaching line of our analysis with a certainty hardly inferior to that of ocular demonstration."

The analysis to which Herschel referred was modern mathematics, as it then existed, applied to the Newtonian law. Mathematically the problem of Uranus was this: given the perturbations, to discover the mass and the orbit of the unknown planet producing them—in short, to discover the unobserved member of the sun's family responsible for the

indisputable disharmony in the otherwise harmonious Newtonian symphony of the solar system.

No simile can convey the difficulty of such a problem to
anyone who has not seen something similar attempted. 'A
needle in a haystack' might be suggested, but here we do
not know whether the haystack exists, or, if it does exist,
in what county it may be, or whether, after all, there is a
needle to be found.

About eleven months before Herschel prophesied, young
Adams had sent to the Astronomer Royal numerical estimates
of the mass and orbit of the undiscovered planet. Moreover,
Adams had calculated where and when the hypothetical
planet perturbing Uranus could be observed in telescopes.
Subsequent events proved his calculations correct within a
reasonable margin of accuracy. Had a patch of sky no larger
than three and a half moons been searched when Adams
told the Astronomer Royal to look, the planet would have
been found. But Adams at the time was only an unknown
quantity of twenty-four. As an undergraduate at Trinity
College, Cambridge, he had resolved to attack the Uranian
problem the moment he was quit of his examinations. On
taking his degree in January, 1843, with the highest honors,
he had immediately set about his self-imposed task.

In the meantime a seasoned mathematical astronomer in
France, Leverrier, had also attacked the perturbations of
Uranus in an attempt to locate and estimate the unknown
perturber. He also succeeded. Adams and Leverrier each
worked in complete ignorance of what the other was doing.

Leverrier was the luckier in his friends. The policy of
'wait and don't see,' immortalized about seventy years
later in World War I by Premier H. H. Asquith (1852–
1928), delayed search for the planet by the English astronomers until Leverrier's livelier Continental friends had al-

ready located the suspect in the heavens—very approximately where both Adams and Leverrier had instructed
practical astronomers to direct their telescopes.

Thus was Neptune discovered by pure mathematical
analysis applied to a great physical hypothesis, and thus
ended one glorious canto of the epic begun by Menaechmus,
carried on by Kepler, and sped well on to its climax—as yet
unpredictable—by Newton.

> Where the statue stood
> Of Newton, with his prism and silent face,
> The marble index of a mind forever
> Voyaging through strange seas of thought alone.
> —*Wordsworth*

> Then felt I like some watcher of the skies
>   When a new planet swims into his ken;
> Or like stout Cortez[1] when with eagle eyes
>   He star'd at the Pacific, and all his men
> Look'd at each other with a wild surmise,
>   Silent, upon a peak in Darien.
> —*Keats*

[1] Keats meant Balboa.

Chapter 14

TWO KINDS OF PICTURE

❖ *14.1* ❖ *Continuity in Science*

Attempts to describe the physical universe have oscillated between two concepts, *continuity* and discreteness. We have glanced at these [4.2], but we must now examine them a little more closely. For clarity I elaborate some of what has already been described.

To our senses, motion is *continuous*. A bullet does not proceed by jerks from point to point of its trajectory, but moves evenly, without breaks in its motion, till it finds its billet. A more familiar conceptual image of continuity is that of all the points on a straight line.

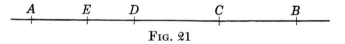

FIG. 21

Imagine any finite segment *AB* of such a line. As we move the pencil from *A* to *B* along this segment, we somewhat vaguely imagine the tracing point passing through 'every point' of the segment without break or interruption. Suppose now that we choose a point at random in the interval from *A* to *B* and label it *C*. Between *A* and *C* we can choose another point, *D*; between *A* and *D*, another point *E*, and so on, 'indefinitely.' Then in each of these smaller segments *AE, ED, DC, CB,* . . . , we can proceed in the same way, getting yet smaller segments, in which the process can again be repeated, and so on, 'indefinitely.' That is, no matter how close together two points of a line may be, we can always imagine a point of the line between those two points. *After* a given point (to the right of it on the line, say) *there is no immediately following point:* there is no point to the right of the given point which can be named the 'next' point on the line.

If this does not seem like common sense, nothing can be done about it, for the concept of 'continuity,' or 'no-nextness' is anything but simple or commonsensical. It is in fact an extremely subtle conceptual abstraction of an ideal situation which, from its very nature, is incapable of sensual experience or experimental realization in the 'actual world,' or in the realm of our 'perceptions.' Nevertheless this highly sophisticated concept of continuity is not repugnant to our intuitions.

We do in fact picture motion as continuous and we do habitually think of the water in a stream as flowing continuously and being itself continuous. We shall see shortly how mathematical analysis takes hold of this problem. If, however, we accept a crude atomism, we analyze the water into its molecules, and instead of the smooth 'continuous' flow of a 'continuous' fluid, we describe the motion of a discrete [4.2] swarm of particles. Each particle will move continuously. We have not abolished *all* the continuity by resolving the water into its individual particles. In the end we must grasp continuity of motion for one particle. If we can do that, all the rest is merely a matter of repetition.

Assuming that we have a firm intuitive grasp of continuous motion, as of a speeding bullet, we can easily see how mathematics is adapted to the description of motion. Some of what follows will reappear in the sequel when we come [Chapter 15] to the calculus. It is of such fundamental importance for the application of mathematics to nature that I may be pardoned a little repetitiousness.

The successive positions of a particle moving continuously are mapped onto, or correlated with, the real numbers in an appropriate segment of the real number system. This system again is a purely conceptual construct, incapable of complete 'realization,' that is, of experimental or visual presentation.

I shall ask the reader to refer here to what was described [4.2] of the real number system. Possibly the description seems trivial. If so, that is only because we have been accustomed to it from early childhood. It was not simple to our ancestors of only four centuries ago. They would have denied that it means anything. The next, in a certain sense an inversion of our intuitions of the real number system, is neither simple nor trivial, even to skilled mathematicians. It is indeed a basic postulate of the mathematics of continuity [4.2, 6.2]. I shall take it in several steps, in an attempt to reveal the sweeping, man-made assumption on which by far the greater part of our applications of mathematics to nature is based.

First, as we saw, there is no difficulty in picturing the rational numbers on the line which already carries images of the integers . . . , -3, -2, -1, 0, 1, 2, 3, ... To any fraction of the form a/b, where a, b are whole numbers (positive or negative, b not zero), we can easily assign as its representative a unique point on the line [4.2].

The last clause in the preceding sentence suggests a difficulty that might easily be subconsciously slurred. As a matter of fact there is no difficulty yet, but there will be in a moment, when we take the last step and plunge headlong into a bottomless infinity of purely conceptual 'numbers.'

Examining the clause in question, a skeptic might ask, "*How* do you assign the representative of a/b?" Similar questions are sometimes met with an embarrassed silence or an appeal to the 'Universal Mind,' as when the grade-school teacher answered an eight-year-old's demand to be told why we subtract when doing long division: "Egbert, there are some things that only God understands." Here the question is referred to Geometry, not God.

In mechanical drawing there is a simple construction for making a length equal to a/b where a, b are any whole numbers, and this construction can be performed with straightedge and compass. Thus the question is disposed of by an appeal to geometrical experience, a depth beneath which we shall not now descend.

Suppose now that it is required to locate the point representative of $\sqrt{2}$ ($= 1.412 \ldots$). Again I refer to [4.2]. The point can still be located with the same implements: construct a square whose side is 1 unit in length; a diagonal of this square is $\sqrt{2}$ units in length. If we seek to construct in a similar manner the cube root of two, our implements are inadequate—as can be proved. Finally, if we wish to locate the point representative of π ($= 3.141592 \ldots$, approximately, the ratio of the circumference of a circle to its diameter) totally new instruments must again be invented. None of those which suffice for square roots, cube roots, and so on, or the roots of *any* algebraic equations with whole-number coefficients, are adequate for the construction of π. I may refer here to what was said about algebraic and transcendental numbers in [11.5] and [11.6].

It is sometimes taken as a reasonable postulate of science that physical 'entities' can be put into one-one correspondence with numbers. Thus it is assumed, for example, that the voltage of an electric current can be expressed as a 'number,' this number designating 'so many times a certain unit of voltage.' A current with an associated voltage of π is easily conceivable, although it could never be measured exactly. So also, in imagination, is a current associated with some other number which no mortal can 'construct'—

in the geometrical sense of producing an instrument which will actually lay down a line segment exactly representing the corresponding number. Moreover it is not very difficult to describe verbally a 'number' of this sort which the entire human race toiling through all its future existence—say for a billion billion years—will never be able to construct.

This brings us to the parting of the ways. Does it mean anything at all to speak of a point correspondent of a 'number' when no human being can perform the construction of such a 'number,' or ever be able to describe how such a construction might be carried out in a finite number of steps?

Some say "Yes," others, "No."

Of the physicists saying "No" are those who since the early 1930s have been driven to metaphysical attempts to understand what is meant by the simplest measurements performed in physical laboratories. These (among them the 'logical positivists') deny meaning to any operation which cannot be performed, at least conceptually, in a finite number of steps. Those saying "Yes" follow Pythagoras in believing that 'number' somehow or other is superhuman and the ultimate alphabet of nature.

Professional mathematicians are divided on this issue into two camps, allied respectively with those of the opposing physicists. The modern Pythagoreans ascribe to all the points on a line unique numerical representatives. Their opponents demand to be shown definite, finitely constructible processes for individualizing the *continuous* infinity ('no-nextness') of *all* the points on any given line segment.

The way out for both mathematicians and physicists is down the same broad highway to what appear to be the insuperable barriers described here in Chapters 19 and 20. Both mathematicians and physicists assume, or postulate, an 'existence' of 'numbers' forever beyond any possibility of human construction or measurement, and all of them develop the consequences of this assumption, the mathematicians in *mathematical analysis* [6.2], the physicists in their *field theories*. Both of these will be sufficiently described in succeeding chapters; for the moment the following is enough.

Mathematical analysis is the body of technical theory based upon the assumption that continuity—'infinite divisibility,' 'no-nextness'—is a self-consistent notion which will not, in ordinary logical argument, lead to contradictions. In field theories it is assumed that the whole of the mathematics of continuity—mathematical analysis, in brief—is logically sound and, when applied to the physical world, will yield only consistent results. It is further assumed that continuity has an image in reality. I shall come back to such questions in the concluding chapter. For the present it is sufficient to have observed that what our intuitions suggest may lead to profound difficulties.

In passing it may be remarked that nobody has yet shown that an erroneous mathematical logic is incapable of predicting phenomena verifiable by experiment. It is possible that exactly the opposite is not only thinkable but is inevitable to the adherents of G. W. F. Hegel (1770–1831) and K. Marx (1818–1883), but we need not go into this here.

If the mathematician assumes, or postulates, that to each point of the line 'corresponds' one 'real number,' and only one, he is faced with this problem: prove that the assumption is self-consistent. To do this he must show among other things that the sum, $a + b$, of any two of the postulated 'real numbers' a,b is again a 'real number'—namely, $a + b$ has a unique 'point-correspondent,' and similarly for the product ab and the quotient a/b. He must show that $a + b = b + a$, $ab = ba$. Instead of attempting anything so difficult (neither a nor b need be 'constructible') he boldly *postulates* the constructibility. The scientist—who as a rule believes in the soundness of the very simplest and therefore most difficult things his mathematical guide does —follows blindly, postulates a self-consistency of *continuous* fields, and proceeds to prophesy, often with complete success. Not one of these prophecies has one whit of validity beyond the human incapacity to construct all the points on the shortest line segment imaginable. No theory which even in the remotest degree appeals to continuity has escaped that drastic, finite, human limitation of being unable to do the things talked about. And, what is more, some mathematicians believe that no theory ever will—unless, of course, human beings surpass their finite limitations and achieve immortality.

Any mathematician or scientist is free to advertise universes of his own creation (and many have), just as any realtor can advertise (and sell) city lots in the middle of the Pacific Ocean so long as he refrains from using the mails for his propaganda. But the moment either the mathematician or the scientist appeals to human powers to validate his check, it bounces back, marked 'no funds.' It is impossible, say the doubters, for any man ever to construct the *continuum* of all the points on a line segment, or ever to depict the concept of *continuous motion* by appeal to the real number system. These things are in essence infinite, beyond constructibility, and of their own nature, superhuman.

Ah, then the superhuman exists! It has been mathematically demonstrated! Has it?

The trap may now be sprung. It catches all those— mathematicians, scientists, philosophers, theologians—who appeal to an 'Infinite Existence' as the ultimate logical justification for their humanly imagined theories. It may be, and indeed seems to be, impossible to prove that any such theory of the infinite can be complete. Beyond this is the problem of proving the consistency of our mathematical systems when pushed out into the infinite. We shall return to this in Chapter 20.

Nevertheless the use of continuity in descriptions of the physical universe continues to justify itself pragmatically in at least two ways: it gives a sufficiently close description of physical phenomena, well within the limits of empirical measurement; it accurately predicts within the same limits. No picture or theory is exact to the last small detail [12.2], and as long as the idealized representation is accepted only as a sufficiently accurate description of what is observable, nobody is misled. Only when temporarily workable hypotheses become creeds is there a risk of believing more than experience justifies.

❖ 14.2 ❖ *Discreteness in Science*

Continuity either is incapable of giving a complete mathematical description of nature or is too cumbrous. A gas, for instance, is pictured as a concourse of individual particles rushing about and colliding according to the laws of mathematical probability. The 'mean free path' between collisions is sufficiently large compared to the dimensions of the particles to isolate the latter as independent individuals; if a gas were treated by the mathematics of continuity, appropriate for fluids, the results would bo too crude for use. All the particles in a gas can be completely enumerated by counting them off 1, 2, 3, ... Such an enumeration for the elements of a continuum is impossible [4.2].

Any set of things whose members can be completely enumerated by counting them off 1, 2, 3, . . . is said to be *discrete* [4.2]. A discrete set does not necessarily consist of only a finite number of elements. The integers 1, 2, 3 . . . themselves are a discrete infinite set. Of (possibly finite) discrete sets in nature may be mentioned the totality of electrons in the universe, or all the stars.

Again, mathematics in its applications deals with much more than physics, and in some of the other domains discreteness is the obvious pattern. Thus if we wish to discuss a given population statistically, we deal from the beginning with a discrete collection. Here, however, if the number of individuals is sufficiently large we approximate to the exact, discrete statistical laws by others based upon continuity.

Mathematical analysis [6.2]—the mathematics of continuity—is far more highly developed than the mathematics of discreteness. A possible reason for this is the greater favor in which continuous pictures of nature were held from the time of Newton to well into the twentieth century. The root of this preference is the assumed continuity of both 'space' and 'time' in terms of which the classical theories of physics were constructed. But the continuity of space and time is a pure assumption. Occasionally some theoretical physicist has struggled to escape from this tyrannical continuity in the hope of emerging into a less cloying space and time. Mathematics offered no hope of benefit from such an escape. The complications of the necessary technique would probably be much worse than what scientists have used since the time of Newton. Decades of purely mathematical work lie ahead of us before anything really useful can be done toward an extensive discrete repainting of the continuous picture. Yet it may have to be done some day if our present theories of atomic structure survive. The description in terms of continuity of situations which are essentially discrete may ultimatly become intolerable.

The most characteristic properties of a discrete set are imaged in the numbers 1, 2, 3, . . . marked at equal intervals along a straight line: it is possible to *order* all the elements of a discrete set so that every element of the set except the last (if there is a last) has a unique successor, and every element except the first (if there is a first) has a unique predecessor; between certain pairs of elements of the ordered set there is no other element of the set, and all the elements of the set can be grouped into such pairs.

Discreteness is the mathematical pattern underlying all theories based on the concept of atomism. All measurements performed in a laboratory or in common life necessarily form a discrete set. Behind motion as we ordinarily conceive it is the mathematics of continuity. At one time in the history of science continuity has been the fashion; at another, discreteness. In the 1920s both became inextricably knotted together in one fruitful confusion. Without abandoning classical logic it is impossible to imagine a third basic pigment, which shall be neither continuous nor discrete, for our pictures of reality.

We must now look in some detail at an actual example of the manner in which continuity enters the descriptions of physical phenomena.

❖ *14.3* ❖ *Eternal Flux*

"Give me matter and motion and I will construct a universe." Such was the modest demand of Descartes. Since the spectacular rise of relativity and the quantum theory mathematical philosophers are inclined to invert Descartes' request, promising to construct matter and motion provided someone will give them the universe. As yet no offers have been made public.

Of the two, matter, with all its attendant abstrusities of nuclear physics, now seems less comprehensible than

motion. It was not always so. Newton's immediate predecessors in the seventeenth century doubtless thought the
ground under their feet solid enough to stand on, while
motion was more or less of an unanalyzed mystery. As the
twentieth century approached its halfway point, matter
seemed more elusive than motion. Indeed, if some of the
major prophets of the physics of the 1940s could be believed,
matter was more like a slightly insane state of mind than
anything else. Motion also, it must be admitted, had dis-
solved in cobwebs of paradox before the searching logic of
philosophical mathematicians. But for science, at least, motion has been less mysterious than matter ever since Newton
and Liebniz developed the calculus. This penetrating instrument of mathematical thought, particularly the differential
calculus, is the natural language of motion and indeed of all
continuous change. It will be described in some detail in the
next chapter; for the present what follows is sufficient.

The intuitive notion of a rate of change presents no
difficulty so long as the rate is constant. Velocity is defined here as rate of change of position, that is, as speed.
(In theoretical mechanics velocity is time-rate of change of
motion in a given direction and in a given sense, positive or
negative. These qualifications are unnecessary for this
discussion. Velocity here has its first dictionary meaning as
speed.) If a car is moving steadily with a velocity of 60
miles an hour, it passes over 88 feet in 1 second, 44 feet in
$\frac{1}{2}$ second, 22 feet in $\frac{1}{4}$ second, and so on. Moreover, the car
advances 22 feet in every $\frac{1}{4}$ second so long as its velocity
remains constantly at 60 miles an hour. (Motorists might
well paste these figures above the speedometer.)

But suppose the brakes are applied, lightly at first, gradually more strongly, till the car stops, say in 15 seconds.
What was the velocity during the 15 seconds? The question

is meaningless. All through the 15 seconds the velocity was changing from instant to instant, and through no second, or tenth, . . . , or millionth, . . . of a second did the velocity remain constant. In fact, no matter how small an interval of time is imagined, provided only that the interval is not zero, the velocity was changing all through that interval. This common problem of everyday life brings us squarely up against the necessity for devising some method which will enable us to reason exactly about continuously varying rates.

To anticipate, it may be said that nearly all the important equations of mathematical physics express relations between rates. One reason why this should be so, historically, is readily suggested if we glance back at Newton's laws of motion [13.3], particularly the first two, and recall that the Newtonian law of universal gravitation concerns the 'force' of attraction between particles. But the second law of motion provides a basis for translating a statement concerning 'forces' into a mathematical equivalent in terms of 'rates.'

Again, if a planet is revolving in its (approximate) ellipse round the sun, the force of attraction between the sun and the planet varies continuously as the planet describes its orbit, the distance between the two bodies varying continuously (and periodically) from a minimum when the planet is at perihelion to a maximum when it is at aphelion. Thus Kepler's problem [13.2] demands variable rates for its mathematical expression.

Dynamical astronomy gave the calculus of rates—*the differential calculus*—one of its strongest initial impulses. After the triumphant success of the Newtonian theory in celestial mechanics [13.4] other departments of physical science were mapped out for conquest by the same general tactics. Again the attack was largely successful, and as late as the early 1900s it was confidently expected that a mechanical theory of the entire physical universe would some day be constructed. Here it is sufficient to indicate on what this abandoned hope was based.

So accustomed are we to picturing the atoms of matter as extremely complicated structures of electrons, neutrons, protons, mesotrons, and possibly further hypothetical -ons, that it is difficult to think ourselves back to the perfectly elastic, unsplittable billiard-ball atoms of nineteenth-century physics. But suppose we do. What more natural, then, than to generalize the Newtonian law to apply to all the atoms in the universe, attracting one another with central forces, but possibly according to laws more complicated than that of the inverse square, depending on the distances between pairs of atoms or molecules? Inverse fifth powers had actually been proposed in certain problems in the kinetic theory of gases.

At the time this general program was not unreasonable. The mathematical difficulties of carrying it through successfully would have been enormous but not necessarily insurmountable. Problems that seem as complicated have been solved by the modern mathematics which began to be used in physics in the 1920s. It was not mathematical obstacles which halted the mechanized description of the universe indefinitely, but a burst of unexpected experimental discoveries, beginning in 1895 with the accidental discovery of X rays and continuing to the late 1930s with the discovery of nuclear fission. But although the grand mechanical program was abandoned, the mathematics which it instigated remained as useful as it had ever been.

❖ 14.4 ❖ Ancient Philosophers and Modern Pedants

Common experience teaches us that continual, varying change is the order of nature. As the old hymn has it,

"Change and decay in all around I see," which seems a needlessly pessimistic paraphrase of the dictum of Heraclitus (sixth century B.C.), "All things flow."

Until mathematics can take a grip on the elusiveness of change, it is powerless to grasp nature in its most characteristic activities. How is this to be done? A moment's reflection will convince anyone that the problem is one of the greatest difficulty. Nevertheless it was solved in the seventeenth century as the flower of a long evolution seeded at least as early as the pre-Socratic Greek philosophers.

If the last seems an excessive estimate of the antiquity of the calculus of rates and of change in general, I recall an experience of D. E. Smith (1860–1944). At the time of his death Smith was the foremost American historian of mathematics. As a young man he went to Germany to study under M. Cantor (1829–1920), at the time the leading world authority in the history of mathematics. Cantor asked the ambitious young man what he would like to investigate. "The history of the calculus." "Well, Mr. Smith, if I were you, I should not go back much farther than Antipho and Bryso"—both of the fifth century B.C. It was not the first time Smith had heard of these primitive Greek analysts. But until Cantor advised him, he had imagined that the calculus began in the seventeenth century A.D. and not some centuries B.C.

What is abstractly the same problem mathematically as that of continuous change in nature occurs almost at the beginning of geometry. A complete and logically satisfactory treatment of the problem of finding the area of a rectangle [12.2] demands a thorough understanding of continuity [4.2]—the very essence of change and flow. It was this problem which introduced the Greeks to the concept of infinite (indefinitely continued) divisibility, continuity, and all the riddles of the mathematical infinite implicit in these intuitively clear notions.

I shall not take space to elaborate this, but I may remind any older readers who learned their elementary geometry in American schools before mathematical instruction was denatured to meet the increasing demands of mediocrity, where they first encountered these fundamental notions. It was in the so-called 'incommensurable case' of all propositions dealing with proportion and areas. Those oldsters who learned their geometry in an English school of the early 1900s followed a way still closer to the ancient Greek tradition as expounded by Euclid in Books V and VI of his *Elements*. Book V, by the way, is usually considered one of the masterpieces of all mathematics, not alone of Greek mathematics. This rigorous introduction to the basic notions of continuity appears to have been replaced in both countries by something easier but inferior logically.

It may be emphasized once for all that an exact description of how mathematics handles the problem of continuous change is neither possible nor desirable in a first approach. Further, to turn aside for a moment if I may be allowed to vent a heretical opinion, I strongly disbelieve in ever giving either engineers or mathematical physicists a rigorous course in the calculus—the mathematics of continuous change. By rigorous I mean going right down to the ambiguous logical roots of the real number system [4.2].

Mathematics is but one of several expedients resorted to by scientists, and in many instances one of the less important. Scientists have checks on their work that no mathematician can ever have. At any stage of an investigation a scientist can compare the results of his calculations with nature itself. To indoctrinate a sanguine young scientist with all the subtle dogmas concerning the meaning of continuity, on which mathematicians themselves see no immediate hope of agreement, is to invite the risk of sterilizing a potentially creative mind. Those self-righteous mathematicians or logicians who carp at what they call the slapdash mathematics of their scientific brothers might more profitably employ their squandered leisure in trying to agree among themselves.

On the other side no theoretical physicist or less reputable user of mathematics has any business whatever in telling a defenseless public what is the human or epistemological significance of any conclusion reached by mathematical reasoning until he has mastered the main tenets of the leading schools of 'mathematical philosophy' since 1900 with all their flagrant doubts and mutual recriminations. Should he attain this mastery, he might lose his gift of tongues for prophesying in the name of mathematics. It would be too much to expect of him that he should cease referring to a mathematics which neither he nor his helpless congregation has had the opportunity to consider and understand critically as a guarantee of the 'truth' of his personal beliefs.

So much for my own beliefs; I shall not obtrude them again. But in passing I thought the reader has a right to know where his author stands on these controversial matters of opinion. They are inherent in all that follows, to the very end of the book.

❖ *14.5* ❖ *Nature in the Small*

To repeat the all-important question in the application of mathematics to nature, how is mathematical reasoning to take hold of the concept of continuous change? The answer is immediate and simple: by mapping the change on the real number system [4.2] as already intimated.

For definiteness consider once more the automobile which is being braked down *continuously* from a speed of

88 feet a second to rest in 15 seconds. We saw that at no instant of the total 15 seconds was the velocity constant. Without attempting to be precise, let us cut up the 15 seconds into a large number of equal time-intervals, say a million, and imagine the motion of the car in any one of these intervals of 15 millionths of a second. Throughout this small time-interval the speed of the car does not change very much. Remembering that speed (velocity), *if uniform,* is measured by the number of units of distance passed over in a unit of time, we now proceed as follows.

Suppose the car travels 450 feet in the 15 seconds. At the beginning of the 15 seconds the car is traveling 88 feet a second; at the end its velocity is zero. We now imagine the 450 feet cut up into so great a number of equal intervals that throughout any one of them the speed of the car does not change very much. (The repetitious phrase is used purposely; compare with the preceding paragraph.)

For definiteness suppose we cut up the distance into a million equal bits. Each will be 450/1,000,000 feet in length, that is, .0054 inch—say approximately, $\frac{1}{200}$ inch. Unless there has been an accident (precluded, because we assumed that the brakes were applied gradually), the velocity of the car will not be *appreciably* different at the end of this $\frac{1}{200}$ inch from what it was at the beginning. We therefore *assume* that: *by taking the distance traveled over sufficiently small, and by assuming that the time for traveling this distance is correspondingly small, we can consider the speed throughout the interval to be UNIFORM, and may calculate it by dividing the fraction of an inch in the length of the (small) distance by the fraction of a second in the (short) time required to traverse the (small) distance.*

Finally, by taking smaller and smaller distances and correspondingly shorter and shorter times, and 'dividing the distance by the time' we reach the concept of *the velocity*

at any given INSTANT throughout the whole 15 seconds
during which the car is being brought to rest, or *decelerated*
from 88 feet a second to zero feet a second.

The methodology of this example is typical of all applications of mathematics to the eternal flux of nature: we translate what is happening in *any sufficiently small region* into

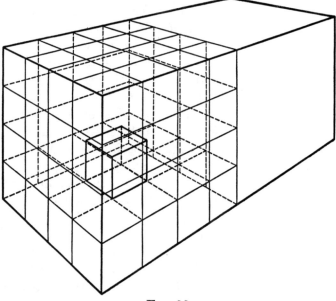

FIG. 22

mathematical symbols, and then proceed to the *limiting*
situation obtained by indefinitely shrinking the dimensions
of the region considered (see Figure 22). We have seen a
purely geometrical instance of this [10.2, 10.3] in discussing
neighborhoods on a surface or, more generally, in any space
in the manner of Riemann.

To take an example, which will be resumed in the following chapter, suppose we are attempting to construct a

mathematical equation describing the flow of the water at *all* points in a river. Near the banks the water dawdles aimlessly, while in midstream it slips purposefully along. Imagine that embedded in the current is an immovable mesh, its threads finer than a spider's silk, of equal cubes. (The argument goes through as well if instead of equal cubes we have equal parallelepipeds, or any equal space-filling volumes.) Select any one of these cubes at random or, more exactly, choose the 'particular general' one. The point at the center of this cube typifies *any* point in the stream. What physical properties does the water have that can be translated into symbols? Very approximately, the water is *incompressible* and, as it flows, no holes appear in it; the fluid is *indestructible*. We wish to express these facts mathematically at 'any' point.

Something more must be assumed. I shall suppose that the velocity of flow is known at each point of the stream. This may seem like assuming all we wish to discover, but it is not so. For example, the assumptions already made are sufficient to determine the streamlines of the flow.

We now attend to the 'particular general'—any—cube we selected and describe what is happening there. By taking a finer and finer mesh, and hence smaller and smaller cubes, and following the description all through the shrinking, we reach in the limit the description of what is happening at the center of the original cube. This is what we wished to know.

But we have not yet stated what it is that we describe in the original cube. When we do it will seem trivial. Note, however, two characteristics which this thing must have: it must refer to the physical data (here incompressibility and indestructibility), and it must be possible to describe it in terms of numbers. The last is necessary as we are seeking a mathematical description, and nearly all the mathematics of classical physics is ultimately expressible as relations between numbers.

Since the water is incompressible and indestructible, the amount of water flowing *into* the cube in one second must be equal to the amount flowing *out*. This is all there is to it. When translated into mathematical symbols as outlined above, this equality of inflow and outflow is expressed in an equation which states the same thing: the *gain* in the amount of water in the cube during one second is *zero*.

This equation involves rates, and is called a *differential equation*, because the study of rates is the object of the *differential calculus*. Over 99 per cent of all the important 'laws' in physical science are subsumed in differential equations, or in systems of such.

❖ 14.6 ❖ *The Scientist Intervenes*

There are two main types of problem in the mathematical investigation of any phenomenon. The first has just been described, the equations (usually differential) corresponding to the situation under investigation must be established— 'set up' is the shop term.

This step is the one that requires scientific insight, not to say genius. For almost any mathematician can write down equations of some sort or other to symbolize even highly complicated physical situations. But unless the equations are amenable to mathematical discipline, of what use are they? This is where Newton and Maxwell excelled. They knew what to symbolize and what to ignore. The average pure mathematician usually produces impeccable and practically useless equations.

❖ 14.7 ❖ *Integration*

The second type of problem posed by nature is inverse to that of setting up the differential equation. The differential equation—that between rates—is to be solved.

In the given equation the rate of change of one variable quantity (say speed) with respect to another (say time) is expressed in terms of other quantities (say those measuring distance and time), and possibly other rates. This is a sufficiently general statement of the situation to bring out the next point of importance.

The *inverse problem* is now to determine from this equation a relation between those quantities whose rates occur in the given equation and the other quantities in that equation. This is called *integrating* the given equation of rates.

Here enters a most important connection between the physical situations pictured by the differential equation and its integrated solution. In the given equation we express the physical conditions *at a particular point at a particular time;* in the solution (integration) of the equation we determine the necessary relation between the variable quantities concerned for *all points at all times.* We thus pass from nature in the small, which we can comprehend and describe, to nature in the large, which we do not know but which we wish to explore mathematically. Again we may note the analogy with differential geometry [10.2].

✤ 14.8 ✤ *Boundary-value Problems*

One further matter must be described before we pass on to a brief recapitulation of all this in easily understood symbols. The complete integrated solution of a differential equation is often so general that it is of but slight use for specific problems. To be useful the generality of the solution must be restricted by conditions which fit the particular problem in hand.

For example, we shall describe in the next chapter the simple equation (differential) expressing the fact that a perfect fluid is incompressible and indestructible. The equation merely translates into mathematical symbols the trivial fact already remarked that as much of a perfect fluid must flow into one cube of a small mesh in unit time as flows out. The equation can be 'solved' quite generally and quite uselessly. If however we impose sufficient conditions on the solution to fit a real problem, such as that the velocity of flow *at the banks* shall be constant, we isolate from all the shapeless infinity of possible flows contained in the 'general' solution a particular one, definitely characterized and unique. In short, by imposing the *boundary conditions* appropriate to definite special problems, we force the general equation to yield the solutions of those problems.

The general equation itself is often no more than a platitude expressed in mathematical symbols. That it should yield valuable information (often unforeseen) when further restricted by limiting conditions—or *boundary values*—is surprising but true.

The fitting of 'general' solutions of differential equations (those involving rates of change) to prescribed initial conditions is coextensive with the theory of '*boundary-value problems*,' and this in turn is a trunk nerve of mathematical physics.

As this matter is really important, I shall briefly outline one more problem relating to it.

There is a simple-looking equation (Fourier's) which expresses the experimentally ascertained law for the flow of heat in any medium. Naturally this equation is a differential equation—one concerning rates. Suppose we wish to determine the temperature at any point of a long uniform rod, one end of which is thrust into a furnace which is known to be at a constant temperature. To make the problem simpler, we shall consider only the situation in which the temperature all along the rod has reached the 'steady state'—no point of the rod is growing either hotter or colder. That this is not an absurd supposition can be felt by sticking the poker into the coals and noting that after a time the end of the poker in the hand grows no hotter. (I have known women who thought that by leaving a flatiron on the stove long enough they could raise its temperature indefinitely.)

What are the 'initial conditions'? Clearly these: the end of the rod in the furnace is at a constant temperature during the whole time; the surface of the rod is radiating heat (according to Newton's law of cooling) into the air at a known rate; at the time considered (and all subsequent times considered) the temperature at any point of the rod is not changing. Our problem is to find a solution of the *general* equation of heat conduction which shall be compatible with these conditions.

To make the problem yet simpler, radiation may be ignored. The simplified problem brings out all the points of methodological importance: solve the *general* equation for the situation considered; make this solution then fit the *special boundary conditions* prescribing the particular problem.

To take a rough analogy, it is something like knowing that all men are mortal and proceeding from that knowledge to calculate the premium which a man aged forty must pay for a stated amount of life insurance. The certainty that a man must die sometime corresponds to the general solution; the 'moral certainty' that he will not see 150 corresponds to the constant temperature of the furnace; the statistical laws of mortality correspond to the general heat equation; and finally the man's age, forty, corresponds to the definite point on the rod whose temperature we wish to know. The equations for heat conduction and mortality are nothing alike.

Chapter 15

THE CHIEF INSTRUMENT OF APPLIED MATHEMATICS

❖ 15.1 ❖ *Rates*

Mathematics as cultivated by professionals contains a great deal that has not yet been applied to science. If we may extrapolate from past experience, much of this vast mass of abstractions will some day find a use in the exploration of nature. Roughly the progress of mathematics and its applications is like a game of leapfrog: at one time pure mathematics is far ahead of any evident uses; at another, scientific advances invite new advances in mathematics.

The chief instrument of applied mathematics is also of the highest importance—in its modern developments—to those who cultivate mathematics for its own sake. This is *the calculus*, the *differential* aspect of which was described in the preceding chapter in connection with velocities; the *integral* aspect will be considered shortly. Here we shall give a little closer description of velocities and of rates of change in general for those who have no phobia for all symbols.

In most of the textbooks the calculus is illustrated by appealing to geometry, particularly to the slopes of tangent lines and the areas bounded by curved lines or surfaces. A reason for this is the historical one that analytic geometry preceded the calculus and did suggest to Fermat a satisfactory calculus of several of the fundamental properties of rates. Newton, however, was more interested in rates as they appear in dynamics, and constructed his calculus accordingly, although he acknowledges that he got the

320

hint of his method "from Fermat's way of drawing tangents." To avoid two difficulties at the same time I shall illustrate the fundamental ideas of the calculus by appealing to our intuitive ideas of motion rather than to the geometry of curves and their tangent lines.

It was in 1694 that Leibniz introduced that most useful word, *function* (or its Latin equivalent), into mathematics. He was concerned with functions of real numbers [4.2]. For ease in following what is coming, I elaborate what was said about functions in [4.2]. Let x,y denote numbers. If x,y are so related that whenever a particular numerical value is assigned to x one value (at least) of y is thereby determined, we say that y is a function of x and write this is $y = f(x)$. The qualification 'at least' in the above may be ignored in what follows. Thus if $y = x^2$, y is the particular function x^2 of x, so that $f(x)$ denotes x^2. If $x + y = 1$ then $y = 1 - x$, so here $f(x)$ is $1 - x$.

A little less precisely, Y is a function of X if, when X is 'known,' Y is also 'known.' The X,Y of greatest use in science are still numbers of some sort; for example, those which measure the coordinates of points on a curve or graph. We shall suppose for simplicity that the x in a function $f(x)$ represents a number of the above sort, that is, until further notice, a *real number* [4.2]. But in the general definition above concerning Y as a function of X, neither Y nor X need be a number of any kind. Such non-numerical functions occur frequently in modern mathematics; for example, as we saw in symbolic logic [5.2], where the X,Y may denote classes, or propositions, or relations.

Another convenient term must be recalled. This is *variable*, already described [4.2, 6.1]. A pertinent example here is the one given in [6.1] of the falling body. The height (measured in appropriate units, as feet, centimeters, and so on) of the body above the earth's surface, is a variable.

From this example we see intuitively what is meant by *a variable approaching a limit*. For the height decreases steadily (continuously) as the body falls, approaching closer and closer to zero height above the earth. Finally the height *attains its limit* (in this case zero) when the body strikes the earth. It is not true that every variable approaching a limit attains that limit.

Next, from our intuitive notion of uniform velocity as a rate, we proceed to generalize the concept of a rate of change so that we can speak of the rate of change of a function $f(x)$ with respect to the variable x.

To make what follows appear reasonable, consider again how we compute the velocity of a particle moving uniformly [14.3]. We divide the number of units of distance passed over in a given time by the number of units of time contained in that time. Thus, if the particle passes over s feet in t seconds, its velocity is s/t feet per second. All this is on the supposition that the particle is moving *uniformly*. Now velocity is '*rate* of change of *distance* (s) with respect to *time* (t),' and we have just computed it, *when uniform*, by 'dividing s by t.'

Consider next how we would proceed to compute the velocity in the general case when it is not necessarily uniform. In the preceding chapter we outlined how this could be done [14.3]; we now translate the description given there into symbols. We shall suppose that the distance s traversed in any time t is a function of t, say $s = f(t)$. This means that for any value T of the time t reckoned after the start of the motion, we can compute the total distance passed over in the elapsed time T by putting T for t in $f(t)$, and calculating out the result. Thus if we are measuring time in seconds and distance in feet, and if $f(t)$ is the particular function t^2 and $T = 3$, we get 3^2, or 9, feet as the distance passed over in 3 seconds.

Given that $s = f(t)$, we are to calculate the velocity at *any* time t. A schematic representation of the process will make the argument clearer. Distances are measured along OX in the direction of the arrow; in t seconds the particle has traversed s feet from O. At a later time, say at the time $t + \Delta t$, the particle will have traversed the distance $s + \Delta s$. In this, Δt is read '*increment of t,*' or simply 'delta t,' and similarly for Δs. Note that Δt is not $\Delta \times t$, as it would be in common algebra, but is merely a shorthand for the words 'increment of t' (see Figure 23).

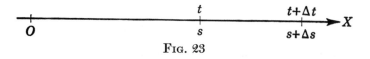

FIG. 23

The length of the interval from s to $s + \Delta s$ is Δs, and this interval has been traversed in the time Δt. If the velocity throughout the interval were uniform, it would be equal to $\Delta s / \Delta t$. But we are not assuming that it is uniform. Suppose, however, that we take Δt so small that Δs is correspondingly small, and the velocity is sensibly (that is, very approximately) uniform throughout this small interval. Then, for this small interval $\Delta s / \Delta t$ is an approximation to the velocity at every point throughout the interval Δs. Taking Δt smaller and smaller, with a corresponding shrinkage in Δs, we get $\Delta s / \Delta t$ as a closer and closer approximation to the velocity at every point in Δs. Finally, if we *pass to the limit*, and let Δt approach zero, Δs also approaching zero simultaneously, we get a limiting value of $\Delta s / \Delta t$ which we call the *velocity at the time t*. The reason we define the velocity thus is evident.

It has been assumed that the limit exists and can be found. When this limit exists it is denoted by $\dfrac{ds}{dt}$, and is called *the derivative of s with respect to t.* Again note that $\dfrac{ds}{dt}$ is *not* the symbol for a fraction, as in algebra, but a shorthand for the limiting value just described.

The actual calculations are carried out as follows. We have $s = f(t)$. Hence when t receives the increment Δt, $f(t)$ becomes $f(t + \Delta t)$. Simultaneously s becomes $s + \Delta s$. Thus

$$s + \Delta s = f(t + \Delta t);$$

whence, since $s = f(t)$, we have

$$\Delta s = f(t + \Delta t) - f(t),$$

and therefore $\dfrac{ds}{dt}$, which is the limiting value of $\dfrac{\Delta s}{\Delta t}$ as Δt approaches zero, is equal to the limiting value of

$$\frac{f(t + \Delta t) - f(t)}{\Delta t}$$

as Δt approaches zero.

An example will show how the work proceeds. Let $f(t)$ be t^2. Going through the steps just outlined, we have $s = t^2$ and find for $\dfrac{ds}{dt}$ in this case the limiting value of

$$\frac{(t + \Delta t)^2 - t^2}{\Delta t}$$

as Δt approaches zero. To continue we simplify the algebra in this expression as far as possible *before* letting Δt approach zero. Thus we get

$$\frac{t^2 + 2t(\Delta t) + (\Delta t)^2 - t^2}{\Delta t}, \; = 2t + \Delta t$$

the *limit* of which as Δt approaches zero is $2t$. Finally then, if $s = t^2$, we get $\dfrac{ds}{dt} = 2t$. As exercises the reader may like to find in this way the derivative of s with respect to t when $s = t^3$, $s = t^4$, $s = t^5$, ... The results are $3t^2$, $4t^3, 5t^4$, . . . , from which the general rule is apparent.

I have not attempted to give a mathematical definition of 'limit,' trusting to common sense to see at least roughly what is meant. A careful statement would require several pages—a truly rigorous treatment would demand a long chapter.

In the above we have found $\dfrac{ds}{dt}$ when $s = f(t)$, and have referred to 'velocity' as an intuitive notion which is easily grasped behind the entire process. But it is clear that if x denotes any variable, and if $y = f(x)$, we might parallel step by step what has been done, and obtain $\dfrac{dy}{dx}$, the *derivative of y with respect to x*, or, in the language of rates, the *rate of change of y with respect to x.*

✤ 15.2 ✤ *Higher Derivatives*

The 'higher' in the heading is a mere technical term which need occasion no apprehension of difficulties to come. Nothing in this section is any more abstruse than anything in the preceding. The simple matter to be explained has been given a separate heading on account of its great importance in scientific applications, particularly in mechanics and other branches of physics. As in defining derivatives, it will be sufficient here to have the intuitive idea of motion at the back of our minds.

A velocity may be uniform or variable. An example of the latter is the velocity of a falling body; the velocity is *accelerated*. In the case of accelerated motion we assume that the *velocity is a function of the time. Acceleration* is defined as the rate of change of velocity with respect to the time. Thus if v denotes the velocity at time t, the acceleration a at time t is the derivative of v with respect to t; that is,

$$a = \frac{dv}{dt}.$$

But v itself is the rate of change of distance (s) with respect to time (t). We saw in fact that

$$v = \frac{ds}{dt}.$$

We therefore have

$$a = \frac{d\left(\frac{ds}{dt}\right)}{dt}.$$

Instead of writing this awkward expression for the acceleration, we abbreviate it to

$$a = \frac{d^2s}{dt^2},$$

which is merely a convenient shorthand for the other.

From what has been said, it follows that acceleration is *a rate of a rate* or, as we say, a *second derivative*. More precisely, acceleration is the second derivative of distance with respect to time.

The acceleration a in its turn may be variable. If so, we can find its derivative with respect to t, namely, $\frac{da}{dt}$. As before we see that this is a *third* rate, and we write it $\frac{d^3s}{dt^3}$.

This process can be continued to fourth, fifth, and so on, derivatives of s with respect to t. *The most important derivatives for applications are the first and second,* namely $\frac{ds}{dt}$ and $\frac{d^2s}{dt^2}$. The first is velocity, the second acceleration.

The third, fourth, and so on, derivatives of distance (s) with respect to time (t) have not received special names.[1]

A reason why these two derivatives should be of physical importance is obvious on referring once more to Newton's three laws of motion [13.3]. These laws are the cornerstone of dynamics. Inspection of them will show that forces are proportional to the accelerations they induce. Thus forces necessarily introduce second derivatives.

All that has been said for s and t here can be paralleled for y and x, where y is a function of any variable x, precisely as in the preceding section.

❖ 15.3 ❖ *Partial Derivatives*

These again are no harder to grasp than the derivatives first defined. They are of supreme importance in science.

So far we have considered functions of only one variable. But nature confronts us with functions of several variables [4.2], and indeed with functions of an infinity of variables. A simple example of a function of two variables is the volume of a gas which depends upon both the temperature and the pressure. It will be sufficient to consider functions $f(x, y, z)$ of three variables x,y,z; the discussion is the same for any finite number of variables.

What happens to $f(x, y, z)$ as x,y,z vary? To attack this we keep all but one of x,y,z fixed (or constant), and calculate the derivative of $f(x, y, z)$ as that one varies. If y,z are held constant while x varies, we get the *partial derivative of $f(x, y, z)$ with respect to x*, and write it $\dfrac{\partial f}{\partial x}$, where f stands for $f(x, y, z)$. We know how to calculate this, for we have already done the like for any function of *one* variable, and in this partial derivative with respect to x, only x is varying

---

[1] In some older works the third derivative is called the *deviation*, but as it is of no great physical significance, the name is not much used.

while we calculate the derivative. In the same way $\dfrac{\partial f}{\partial y}, \dfrac{\partial f}{\partial z}$ may be calculated, the first by holding x,z constant while y varies, the second by holding x,y constant while z varies.

The symbol ∂ of partial derivatives is not a letter from any alphabet but a specially invented mathematical sign. It may be read 'partial.'

Precisely as in the preceding section we can proceed to *higher* derivatives of this kind. Thus, still keeping y,z constant and letting x vary, we can calculate the partial derivative of $\dfrac{\partial f}{\partial x}$ with respect to x. The result is called *the second partial derivative with respect to x*, and is denoted by $\dfrac{\partial^2 f}{\partial x^2}$. Similarly for $\dfrac{\partial^2 f}{\partial y^2}, \dfrac{\partial^2 f}{\partial z^2}$.

Or we may start with $\dfrac{\partial f}{\partial x}$, and in it keep x,z constant, let y vary, and calculate the derivative with respect to y. This is written $\dfrac{\partial^2 f}{\partial y\, \partial x}$. By what has just been said it is by no means obvious that

$$\frac{\partial^2 f}{\partial y\, \partial x} = \frac{\partial^2 f}{\partial x\, \partial y}.$$

On the right we hold y,z constant while calculating the derivative with respect to x of $\dfrac{\partial f}{\partial y}$. Yet the equation written is true—provided all the derivatives actually can be calculated ('exist' is the more correct technical term). More precisely, if $\dfrac{\partial f}{\partial x}, \dfrac{\partial f}{\partial y}$, and $\dfrac{\partial f}{\partial z}$ all exist in the neighborhood of the point (x, y), and if $\dfrac{\partial^2 f}{\partial x\, \partial y}$ is continuous at (x, y), then $\dfrac{\partial^2 f}{\partial y\, \partial x}$ exists at (x, y) and these two second partial derivatives are equal. For most actual physical problems of more than academic interest, the preceding requirements are satisfied.

❖ *15.4* ❖ *Differential Equations*

An equation containing derivatives is called a *differential equation;* if the derivatives are partial (∂), the equation is a *partial differential equation*. The most interesting equations of mathematical physics are of the latter kind. One very famous equation of this sort is Laplace's,

$$\frac{\partial^2 u}{\partial x^2} + \frac{\partial^2 u}{\partial y^2} + \frac{\partial^2 u}{\partial z^2} = 0,$$

which appeared first in the Newtonian theory of gravitational attraction. It occurs also in the theories of elasticity, sound, light, heat, electromagnetism, and fluid motion. We shall indicate in a moment how this equation turns up in hydrodynamics; for the present we must state what is done with such an equation when we have it.

The 'unknown' in the above equation is u, and we are required, given the equation, to find u as a function of x,y,z. This is called *integrating* the equation. It need hardly be said that such a problem is much harder than solving an algebraic equation, even of very high degree, numerically. The 'most general' u satisfying the equation would not be of much use physically; what is wanted is *a particular u that satisfies certain given conditions*. In other words we are to solve a *boundary-value problem* as described [14.8] for the example of the heated rod. The art of solving such problems of the sort that do occur in science is a difficult one. Sometimes the solution of a new physical problem leading to a boundary-value problem demands the invention of new branches of mathematics, or at least the creation and investigation of new kinds of functions. When the solution has been obtained we usually derive much information about the physical problem that we did not consciously put into the equation when we constructed it to fit the problem.

✢ 15.5 ✢ *Fluid Flow*

Returning for a moment to Laplace's equation, I shall briefly indicate what it signifies in hydrodynamics. There are two distinct ways in which the problem of giving a mathematical description of the motion of a perfect fluid may be attacked. In the first we fix our attention on *any* particle of the fluid, meaning by 'any' the 'particular-general,' and construct the equations of motion of this particle. In the second method we attend to *any* mesh of the network imagined (as in [14.5]) in the fluid, and observe what is happening in that mesh. In the first we explore the whole fluid by traveling with the typical particle; in the second we stand still at the typical point and watch the fluid stream by. The 'elementary mesh' is ultimately to be taken so small that we can neglect powers higher than the first of the lengths of its sides. (If a side is .001 inch long, the square of this is .000001 inch long, which is 'negligible' in comparison with .001.) We shall consider the second method.

In Figure 24 three mutually perpendicular planes have the origin O in common. The three systems of lines composing the network are parallel respectively to these three planes; a single mesh of the network is represented in the figure. The point P at the center of this mesh has the coordinates (x, y, z); that is to say that the distance of the point P from the plane YOZ is x, from the plane XOZ the distance of P is y, and from the plane XOY the distance is z. Let the length of AB be $2\Delta x$, that of BC, $2\Delta y$, that of CF, $2\Delta z$. For the meaning of Δ, see [15.1]. Then the coordinates of A are $(x - \Delta x, y, z)$, those of B are $(x + \Delta x, y, z)$, and similarly we can write down the coordinates of the remaining vertices. Our problem is to express mathematically the fact that a 'perfect fluid' is incompressible and indestructible, or, in less mystical language as we noted [14.5], that as much of the fluid flows out of the mesh in a unit of time as flows into it. The flow is assumed to be *continuous* so that no holes or bubbles appear in the fluid.

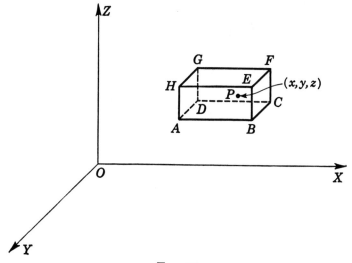

Fig. 24

The velocity of flow at any point can be compounded from the three velocities parallel respectively to OX, OY, OZ. The total amount flowing into the mesh in one second is obtained by adding the flows in these three directions. The velocity of flow at any point is a function of the coordinates of the point. Say the three component velocities at any point (x, y, z) are U, V, W, where U stands for $U(x, y, z,)$ etc. Then across the face $ADGH$ the velocity is $U(x - \Delta x, y, z)$; across the opposite face it is $U(x + \Delta x, y, z)$. To a first approximation (which become increasingly exact as the mesh is taken finer and finer) these can easily be shown to be

$$U - \Delta x \frac{\partial U}{\partial x}, \; U + \Delta x \frac{\partial U}{\partial x}.$$

Hence, if D is the density of the fluid, which we shall assume to be constant, the total mass of fluid entering at the first face in unit time is

$$D\left(U - \Delta x \frac{\partial U}{\partial x}\right)\Delta y \, \Delta z;$$

the total mass flowing out across the opposite face is

$$D\left(U + \Delta x \frac{\partial U}{\partial x}\right)\Delta y \, \Delta z.$$

The excess of the first over the second is

$$-2D \frac{\partial U}{\partial x} \Delta x \, \Delta y \, \Delta z.$$

In exactly the same way, if we consider the flows in the directions OY, OZ, we get

$$-2D \frac{\partial V}{\partial y} \Delta x \, \Delta y \, \Delta z, \; -2D \frac{\partial W}{\partial z} \Delta x \, \Delta y \, \Delta z.$$

The sum of all three of these excesses is the total excess of inflow over outflow. But the fluid is 'perfect,' so that this excess is zero. Canceling out $-2D \, \Delta x \, \Delta y \, \Delta z$ from the sum equated to zero we get

$$\frac{\partial U}{\partial x} + \frac{\partial V}{\partial y} + \frac{\partial W}{\partial z} = 0.$$

From this partial differential equation connecting U, V, W we should be required to find the relation connecting x, y, z

which is satisfied for given boundary conditions. The equation contains *three* 'unknown' functions, U, V, W; it would be much simpler mathematically if it contained only *one*. The idea occurred to Lagrange (and after him also to Laplace) of postulating that U, V, W are respectively the partial derivatives with respect to x, y, z of *the same function*, say u, so that

$$U = \frac{\partial u}{\partial x}, \ V = \frac{\partial u}{\partial y}, \ W = \frac{\partial u}{\partial z}.$$

(It is customary to take the negatives of these, but this is of no consequence here.) Making this change we get *Laplace's equation* (in three dimensions)

$$\frac{\partial^2 u}{\partial x^2} + \frac{\partial^2 u}{\partial y^2} + \frac{\partial^2 u}{\partial z^2} = 0.$$

Glancing back over the argument we see that this equation expresses the fact (which we assumed) that the fluid is incompressible and indestructible. The function u is called a *velocity potential*. If this *potential function* exists, as we have supposed, the motion is said to be *irrotational*, that is, there are no vortices in the flow. The last is merely stated and is not meant to be evident, for it is not.

Now, we might not be able to deduce much by common reasoning from the fact that a fluid is perfect and its motion irrotational, together with given boundary conditions, such as that the velocity is constant along certain surfaces. In fact, we seem to have been translating platitudes into symbols. It is all the more astonishing, then, that from platitudinous assumptions such as those in this and other physical problems, which fit observations closely enough for all scientific and practical purposes, we should be able by the techniques of mathematics to deduce results which are anything but platitudinous. As an example I may refer to Maxwell's work [12.1] in electromagnetism. All I have attempted here is to suggest how physical trivialities are translated into mathematical symbols.

It seems clear that if we are to get general results from our mathematics the postulates from which we start must be simple. The greater the number of restrictions imposed at the beginning, the narrower will be the scope of our deductions. After all, then, it is not surprising that the really important equations of mathematical physics are translations into symbols of empirical postulates of the extremest simplicity. The solving of equations satisfying given boundary conditions is seldom simple and is a task for the skilled mathematician. It cannot be learned in a day. If you want to do this sort of thing as a lifework, start at sixteen or seventeen, not later than nineteen.

❖ 15.6 ❖ *Integrals*

The calculus commonly means the union of the *differential* calculus and the *integral* calculus. The differential calculus, which we have just described, is concerned with rates or derivatives. The integral calculus has two aspects, the first of which is *the inverse problem of rates;* that is, given a derivative, to find at least one function which produces that derivative. The second aspect deals with *the computation of a certain special type of limiting sums,* to be described presently, and for this reason the integral calculus is sometimes called the *sum* calculus (although the term is also used in the *calculus of finite differences,* which I shall not describe). This second aspect of the integral calculus goes back at least to Archimedes, although it was more or less lost sight of until the seventeenth century, when Newton and Leibniz developed *the* calculus.

Roughly, a capital use of the integral calculus is the solution of problems translated from nature into the language of the differential calculus. The two calculuses supplement each other, and a tolerably satisfying description of the eternal flux of nature would hardly be attainable without both. Although on the purely technical side the integral calculus is much harder than the differential, its objectives are much easier to grasp intuitively. This may explain why the integral calculus so long preceded the differential. Neither Archimedes nor any of his successors until the great age of Newton and Leibniz had sufficient algebra and analysis to comprehend, much less to overcome, the technical difficulties inseparable from the actual practice of integration. These difficulties do not concern us here; the simple idea generating them does. It may be said that only a very skillful analyst could propose a problem in the diffierential calculus which some wilier analyst could not solve. But any tyro who has spent a few hours mastering the meaning of the symbols used in the integral calculus can manufacture problems which no mathematician living can solve and which, so far as we dare prophesy, no mathematician will solve in the next hundred years. I would write down one such here were I not afraid that by ill luck I should happen to pitch on one which any sophomore in college can do. Nevertheless, I believe that any professional will endorse the spirit of my remarks. If not, I accept the challenge of composing a simple problem that will drive all doubters to cover. That it would be a fantastically useless problem is beside the point.

It is not a question of grinding out successive approximations by hand or by modern calculating machines. Calculation can produce a number that may look reasonable but may be illusory or nonsensical. Mathematicians may be reminded of improper integrals which the calculator or the gullible machine swallows and regurgitates in yards of meaningless digits. And no machine yet constructed has had brains enough to decide whether the sequence of numbers it extrudes converges or is asymptotic (in the usual technical sense) to the value of the unknown sought.

Let us give the machine and the stubborn calculator their dues. The machine can calculate tide tables, mortality tables, the trajectories of guided missiles, and the like, more expeditiously than any human being could do the same. But ask the machine whether triple collision in the problem of three bodies mutually attracting one another according to the Newtonian law is possible. The mechanical answer is an engimatic table of numbers. Again, it is true that digit by digit the 'solution' of a non-linear differential equation can be calculated. Unless it is proved or is evident that the calculations are not divergent, and therefore meaningless, they are worthless. It is the part of mathematics in such calculations to show that we are not merely wasting our labor.

The main concept to be grasped is that of a *definite integral*. The 'definite' is part of the technical definition. An *indefinite integral* is the *inverse of a derivative* (as will be explained). A definite integral is simply a limiting sum of the kind mentioned. The mathematical miracle is that these two kinds of integrals should be related in a simple and beautiful manner, which we shall note presently, in the *fundamental theorem of the calculus*. The notion of a definite integral is intuitively obvious when presented graphically, as I shall do.

A practical problem occurring frequently in science and technology is to find the area bounded by specified straight or curved lines. The key problem to this is the following: find the area $ABB'A'$, where AA', AB, BB' are finite segments of straight lines, AA' and BB' are perpendicular to AB, and $A'B'$ is a specified continuous curve without discontinuities (roughly, breaks or sudden jumps) anywhere from A' to B' (see Figure 25).

A child could tell us how to go about solving this, precisely as Archimedes did—up to a certain stage. In fact, children of ten have almost instantly got at the gist of it all when a pair of scissors was laid within reach. A child (or Archimedes) leading us, we cut the area $ABB'A'$ into any convenient number of strips of equal breadth, snip off the

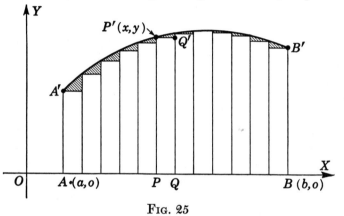

FIG. 25

awkward shaded curved bits at the top of the strips by the shortest cuts possible (perpendicular to the sides of the strips), and get an assortment of rectangles. The area of a rectangle is given by the rule 'length times breadth.' Adding the areas of all the rectangles we have an approximation to the required area $ABB'A'$. The thinner the strips the better the approximation.

The problem of finding an exact measure of the area $ABB'A'$ reduces to this: find first a formula for the sum of the areas of the rectangles when AB is divided into n equal bits; find then the *limit* of this sum *as n increases indefinitely*, or, as we say, when n *approaches infinity*, or *tends to infinity*.

The *greater* n is, the *less* will be the error committed by discarding the shaded bits. It is intuitively evident—and can be proved—that as n approaches infinity the sum of the discarded bits approaches zero as a limit. Hence, *in the limit*, the sum of the rectangles will be the required area.

The first step is easily taken when (as in analytic geometry, Chapter 7), we know the equation of the curve of which $A'B'$ is an arc. Say the equation of this curve is $y = f(x)$, and let (x, y) be the coordinates of any point on the arc $A'B'$. Let the coordinates of A be $(a, 0)$, those of B, $(b, 0)$. Then the length of BA is $b - a$, and hence one-nth of this length is $(b - a)/n$. If BA is divided into n equal bits, $(b - a)/n$ is the breadth of each strip in the figure.

Look now at the typical strip $PQQ'P'$. Its breadth is $(b - a)/n$; its side PP' is $f(x)$, since $y = f(x)$ is the equation of the curve, and $OP = x$; its side QQ' is $f[x + (b - a)/n]$, since $AQ = AP + (b - a)/n$. To approximate to the area of this strip we take PP' as the height of the rectangle after the top shaded bit has been discarded. Thus the *approximate* area of the strip is $PP' \times PQ$, that is, $f(x) \times (b - a)/n$; the *next* strip would have as its approximate area

$$f\left(x + \frac{b - a}{n}\right) \times \frac{b - a}{n};$$

the next to this,

$$f\left(x + 2\left(\frac{b - a}{n}\right)\right) \times \frac{b - a}{n};$$

and so on. The *first* strip is approximated by $f(a) \times (b - a)/n$, since AA' is $f(a)$ units in length; the *last* strip has the approximate area

$$f[a + (n - 1)]\left(\frac{b - a}{n}\right) \times \frac{b - a}{n}.$$

The approximation to the *total* area $ABB'A'$ when AB is cut into n equal bits is therefore the *sum* of all expressions of the form

$$f\left[a + s\left(\frac{b-a}{n}\right)\right] \times \frac{b-a}{n}$$

that we get by *adding all these* for $s = 0, 1, 2, \ldots, n - 1$. This almost concludes the first step. To finish it the algebra of the indicated addition is performed. This gives *that approximation* to the required area which is obtained by cutting AB into n equal bits. This will yield a formula involving n. The second step is to see what this formula ultimately becomes as n increases indefinitely.

The last is not necessarily as hard as it sounds. Frequently we get expressions like a/n, a/n^2, and so on, whose limits are required as n 'approaches infinity.' These limits are (intuitively) zero, as 'the larger the denominator, the smaller the fraction.' Consider, for example, the limit of the sequence $\frac{1}{10}, \frac{1}{100}, \frac{1}{1000}, \frac{1}{10,000}, \ldots$

This is about as far as a child could get, or as Archimedes actually got, except in a few very simple problems. The child of course would use no such barbarous symbols as we have used, but would describe in syllables of limpid simplicity exactly the same limiting process as that which the symbols express.

I now transcribe the above description into the customary mathematical symbolism. Recall that the *equation* of the curve under consideration is $y = f(x)$; that A has the coordinates $(a,0)$, and B, $(b,0)$, so that the area $ABB'A'$ could be generated by AA' sweeping it out as x [in $y = f(x)$] varies from the *lower limit* a to the *upper limit* b; and recall finally that the area $ABB'A'$ is obtained as a limiting *sum*. Then it is graphic to write for this area the expression

$$\int_a^b f(x)\ dx.$$

For, *dx in the limit*, when n is increased indefinitely, signifies the *breadth* of the typical rectangle, $f(x)$ is the height of this rectangle; $\int$ is the old-fashioned S, signifying *sum;* a, the *lower* limit is placed *below* and b, the *upper* limit, *above.* Never was a more expressive mathematical symbolism devised.

The symbol $\int_a^b f(x)\ dx$ is called the *definite integral with respect to x from a to b of the function f(x)*.

As what has just been described is one of the fundamental ideas of all mathematics—and an indispensable tool in the application of mathematics to nature—it will repay the reader to look over what has been done once more or, much better, to get hold of a good introductory book on the calculus and really master the technique for calculating integrals. A sufficient mastery can be gained in six school months by any boy or girl of sixteen or seventeen. The result of acquiring a *reading knowledge*, which is much less than a *working knowledge*, of this particular symbolism is like that of simultaneously mastering a reading knowledge of a dozen languages. Many of the great literatures of science which hitherto were arcana become clear as daylight once the calculus is a familiar script, and the reader is no longer compelled to accept secondhand—and often obscure—translations of the originals. A critical insight into modern theoretical physics, such as is necessary for any philosopher of physical science, is possible only to one who can read and understand the facile symbolism of the calculus.

❖ *15.7* ❖ *The Fundamental Theorem of the Calculus*

The above is the title by which the theorem next described is known. The theorem connects the differential calculus and the integral calculus. It also short-circuits the labor demanded for the calculation of limiting sums expres-

sing definite integrals. I can only state the theorem, referring the reader to any text on the calculus for a proof. To
calculate

$$\int_a^b f(x)\ dx$$

we proceed as follows. We suppose the $f(x)$ is finite and
continuous for all values of x from a to b.

(1) Find the function of x, say $F(x)$, whose derivative
with respect to x is equal to $f(x)$. That is, find $F(x)$ such that

$$\frac{dF(x)}{dx} = f(x).$$

(2) Calculate the value of $F(x)$ when $x = a$ and when
$x = b$, getting $F(a)$ and $F(b)$. Then

$$\int_a^b f(x)\ dx = F(b) - F(a).$$

This, omitting refinements, is the whole story. The difficulties occur at the first step. However, after incessant
work on the calculus since the seventeenth century, mathematicians can either find the $F(x)$ required [when $f(x)$ is
given] exactly, or they can approximate to $F(a), F(b)$ (as in
the second step) to any desired accuracy, for those definite
integrals which are of scientific importance.

Chapter 16

FURTHER CALCULUSES

❖ 16.1 ❖ *Greatest or Least*

There is yet another calculus of great use in the sciences, *the calculus of variations*. It originated in the seventeenth century with problems of maxima and minima, and today in its purely mathematical aspects is so extensive that much of a working lifetime might be required for its mastery.

In many problems it is important to know under what conditions some variable number (measuring distance, or energy, or action, say) assumes a greatest *or* a least value— that is, a value which is 'greatest' or 'least.' Graphically these *maxima* or *minima* would correspond to the points $A, B, C, \ldots, I, J$ (see Figure 26).

For convenience 'a greatest,' 'a least' are included in the phrase 'an *extremum*.'

This description is sufficiently close for an account like the present. But like all statements in a merely descriptive paraphrase, this one omits certain exceptional cases that may not meet the eye on a diagram. These exceptions, however, are usually of only minor importance in applications to science, and were not even noticed until long after the unsophisticated form of the theory of extrema had been used with great success in physics. This does not mean that mathematical refinements can be safely ignored, even in science. They cannot. So recently as 1935 a spirited controversy developed over the exact physical meaning of Fermat's principle (to be stated presently). A usual form of the principle was shown to be inadequate and misleading in certain freak situations in optics. Such situations would arise only by a sort of miracle, but the fact that they can occur at all must be taken account of in a scientific discussion. Here we shall attempt only to give a rough, intuitive account of the way in which problems of extrema enter science.

It sounds like common sense to us, as it did to Newton, to say that a variable is increasing when its rate of change is positive, decreasing when its rate of change is negative,

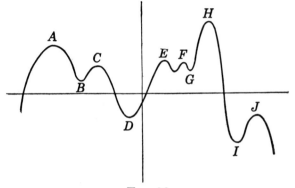

Fig. 26

and neither increasing nor decreasing, that is, *stationary*, when the variable is at an extremum. Hence at an extremum the rate is zero. On the diagram this means that the tangent line to the curve is horizontal at a maximum or minimum.

Referring to what was said about rates [15.1] we therefore have the following method for locating extremal—more generally, *stationary*—values of a function. A stationary value which is neither a maximum nor a minimum is shown at *S*. If the variable y is a function of the variable x, say $y = f(x)$, the values of x which make y stationary will be found by equating to zero the derivative of y with respect to x, and solving the resulting equation for x. For this derivative is the rate of change of y with respect to x, and at a stationary value of y, as we have just seen, this rate is zero (see Figures 26 and 27).

As *least* values (minima) have been of the first importance in the physical sciences, we shall attend to them alone, understanding, however, that a fuller account would be based on extrema.

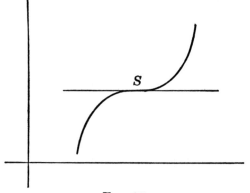

Fig. 27

The importance of minima in physics can be imagined from the earliest successful use of them in science, Fermat's 'principle of least time,' enunciated in 1662. Fermat undertook to deduce the laws of optics as known in his day from a single principle. These laws include those of reflection and refraction; for example, the angle of incidence is equal to the angle of reflection (see Figure 28). Fermat showed that the path of a ray of light traversing any medium from one fixed point to another is such that the time taken for the actual path is less than would be the time required for any other path joining the two points (in reflection the path must go via the mirror).

This principle is typical of many. Often the whole of a complicated department of mathematical physics is summed up in the statement of a corresponding 'least' or, more accurately, 'stationary,' principle, from which the fundamental equations of the subject can be deduced by pure mathematics. Such is the case, for example, in mechanics, optics, relativity, and much of electricity and magnetism.

Fermat's principle suggests a very definite mathematical problem, the key to the whole situation and all like it. A

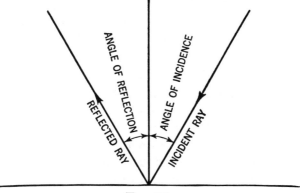

ANGLE OF REFLECTION
ANGLE OF INCIDENCE
REFLECTED RAY
INCIDENT RAY

FIG. 28

ray of light traversing a material medium of variable density will pursue a path which may be as twisted as we please, provided the density of the medium varies from point to point in a sufficiently complicated manner. Along such a path the velocity of the light will vary from point to point. If the velocity were constant for the whole path, and if the path were a straight line, there would be no difficulty in calculating the total time of traversing the path. But in a medium of variable density this simplicity is absent. We proceed as we did [14.5, 15.1] in calculating a variable velocity, or a variable rate: split the path up into equal small bits and approximate to the correspondingly small time consumed in traversing the typical small bit.

Here, however, we are faced with a new difficulty. All these small times must be *added*, the *limit of the sum* must be found as the division points are taken closer and closer together on the path, and finally the *least* value of this *limiting sum* must be calculated. Limiting sums of the kind just described are the *definite integrals* already discussed [15.6]. The new difficulty enters with the problem of *minimizing* a definite integral. An example from mechanics will illustrate the most celebrated of all external principles, that of *least action*. Incidentally this will illustrate another application of the geometry of manifolds [10.2].

❖ *16.2* ❖ *Least Action*

Momentum in mechanics (as 'motion' in Newton's second law, [13.3]) is 'mass multiplied by velocity.' Suppose a mechanical system changes from one state—roughly, position—or *configuration*, to another. We shall assume that no energy is dissipated. For definiteness we may think of the system as a swarm of particles subject to the laws of ordinary (Newtonian) mechanics. Each particle in the swarm will be dynamically specified by giving its position coordinates x,y,z as functions of the time t, and its momentum coordinates u,v,w, also as functions of t. The latter are the three components of momentum of the particle in directions parallel respectively to the axes along which x,y,z are measured. Thus the particle has *six* coordinates, and may be specified by the sextuple (x, y, z, u, v, w) giving its position and momentum at any time t.

Suppose there are precisely n particles in the swarm. Each has 6 coordinates: the whole swarm has $6n$. Thus, considering the swarm as a single thing, we may regard it as being represented by a 'point' in a 'space' of $6n$ dimensions (compare with [10.1]). As the swarm passes from one configuration to another its 'representative point' in the space will trace out a curve. If this is difficult to imagine, think only of two dimensions; the mere language of geometry does the rest.

What sort of curve will the representative point of the swarm follow from one position A in the 'space' to another, B, as the time changes? The answer is similar to that for Fermat's optical problem. Here, however, instead of the *time* it is the *action* which is *least*. As before the path is split up into equal small bits, the action in each is approximated, and the limiting sum is taken; that is, the action is *integrated* along the path. *Action* is measured by 'momen-

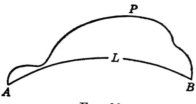

Fig. 29

tum times distance.' This is a reasonable concept, for the number of equally spaced panes of glass a heaved brick can smash depends upon both the mass and the velocity of the brick, and the total 'action' of the brick can be judged from the number of broken panes. Any normal boy has known this since he was five.

Schematically the 'principle of least action' just described can be represented as follows. If A,B represent any two points in the 'configuration space'—the manifold of $6n$ dimensions corresponding to the swarm—ALB represents that path along which the total action is least. Then ALB represents the *actual* path followed, while along any other conceivable path, say APB, the action would not be the least and this path is not followed (see Figure 29).

This principle was stated in 1747 by P. L. M. de Maupertuis (1698–1759). Its niggardliness caused Maupertuis to exclaim at the divine economy, not to say the parsinomy, of God for having so created the universe that nothing in it, not even 'action,' is squandered.

If the motion is constrained to be along a geodesic in the space concerned, the action need not be the *least*, as we shall see at once. The following example may suggest the desirability for caution in making wide philosophical or theological generalizations from 'least action' or any other principle of mathematicized science. This is particularly germane as it relates to a curved space. The 'space' of relativity [10.3] is 'curved.' The example is due essentially to Jacobi, one of the architects of modern dynamics.

Suppose the earth were a smooth globe. Imagine a small, smooth sphere resting on this ideal earth at the spot A and held there by the tug of gravity. Under the Newtonian law of attraction [13.3] the small sphere can move freely only along a geodesic on the smooth earth. The geodesics [3.3] are arcs of great circles. Meteors are shooting about all through the circumambient atmosphere. One of these hits the sphere resting at A. Under the force of gravity combined with the impulse imparted by the meteor, the sphere moves from A to another point B on the globe. We are assuming the impulse to be sufficiently small not to knock the sphere clean off the globe. Provided the sphere is hit appropriately, the track from A to B will be an arc of a great circle. Maupertuis would say that the sphere will travel from A to B by the *shortest* arc of the great circle joining A and B. Unless A and B are the extremities of a diameter of the sphere, when the path from A to B is equal to that from B to A, one of these paths will be the *shortest* consistent with the mechanics. This is true for one possible hit. If the meteor had happened to hit the sphere on the opposite side, the path from A to B would have been the *longest*.

So much for the divine economy of nature as manifested in the principle of 'least' action. And so much also for all extrascientific speculations based on the metaphors of mathematical science.

❖ *16.3* ❖ *The Calculus of Variations*

This is the technical name for the mathematics of minimizing (more precisely, finding extrema of) definite integrals. In the problem of a swarm of particles chosen to illustrate 'least action' we had an instance of this calculus. There (see Figure 29) we had to find the path ALB along which the total action of the system in passing from the configuration represented by A to that represented by B is less than the total action along any other path. Combining this remark with the other [16.1], that at an extremum the rate of change is zero, we see that the following is reasonable: if the path ALB is *slightly* distorted, the variation in the total action must be zero. There are techniques for calculating the variation. The result of equating the variation to zero in any reasonable physical problem automatically produces the differential equations [15.4] of the problem, and this is sometimes the simplest way of deriving them. But there must be a 'minimum principle' on which to operate in the first place. There have been attempts to extend the variational technique to human ecology, as in G. K. Zipf's (1948) 'principle of least effort'—solved daily without mathematics by the lazy.

❖ *16.4* ❖ *Hamilton's Prediction*

To conclude this very slight sketch of variational principles I recall one of the most spectacular predictions in the history of science, that made in 1832 by Hamilton. By a brilliant application of his variational principle in optics,

Hamilton predicted a totally unsuspected optical phenomenon and moreover stated its numerical magnitude in advance of the experiments which verified the prediction.
What follows is based on Hamilton's own account of his
discovery of *conical refraction*.

In one of his philosophies of optics, Hamilton considered
that science as the geometry of a manifold of eight dimensions [8.3]. Of the eight coordinates, six were the ordinary
Cartesian coordinates of two variable points in common
(Euclidean) space of three dimensions, the seventh was an
index of color, and the eighth what Hamilton called the
characteristic function. Speaking of the last, Hamilton
says, "In the manner of its dependence on the seven first
[coordinates] are involved all the properties of the [optical]
system." By analogy with Maupertuis' principle, Hamilton called his function the *action* between the two variable
points. Then, by variation of the characteristic function
corresponding to any small variations in the positions on
which it depends, he obtained "a fundamental formula"
which, he stated with justifiable enthusiasm, was equivalent
to all problems in mathematical optics, "respecting all
imaginable combinations of mirrors, lenses, crystals, and
atmospheres."

By varying the characteristic function according to the
technical processes of the calculus of variations, Hamilton
was thus enabled to obtain the differential equations corresponding to any conceivable problem in optics. *He made
a similar advance in mechanics by the same means.*

A strong point of Hamilton's method in optics is its
independence of hypotheses concerning the nature of light.
His principle is therefore immune to the decay which sooner
or later seems to overtake all scientific theories.

In crucial support of his claims Hamilton predicted conical refraction. It had long been a fact of observation that

certain crystals exhibit double refraction—a ray of light is not only bent on passing through the crystal but is split into two, both bent. But no more than two such refracted rays had ever been observed or even suspected. Hamilton proved that there ought to be, in certain cases which he designated, not only two, three, four, . . . but *an infinity*, or *cone*, of refracted rays *within* the crystal corresponding to and resulting from a *single* incident ray; in other cases a single ray within the crystal should give an emergent cone. These are practically his own words.

The cones were looked for and, after a delicate experiment, found. The predicted angles of the cones also agreed with experiment.

❖ 16.5 ❖ *The Complex Variable*

A distinctive feature of progress in analysis during the nineteenth century was the stupendous development of the theory of functions of a complex variable. As it is now a common tool for workers in the physical sciences and some of the technologies such as aerodynamics, I shall pass this topic with a mere mention. It is a comparative commonplace of mathematical instruction in the better technical schools today; in the early 1900s it was reserved for dreary lectures to graduate students in pure mathematics.

We noted [4.3] that if all the postulates of common algebra are retained, then no numbers more general than the complex [4.2] satisfy the postulates. This gives a strong hint why functions of a complex variable sweep up so much of analysis. The development had well started by 1830; in fact Cauchy had then made his greatest contributions to this branch of analysis, of which he was the creator.

Later in the century two other ways of looking at the whole subject were discovered, one by Weierstrass, the other by Riemann.

Weierstrass *arithmetized* analysis. His universal tool was the *power series*. He regarded functions from the point of view of the convergent infinite series like $a_0 + a_1z + a_2z^2 + \cdots + a_nz^n + \ldots$, which define the functions for values of the variable z in ranges appropriate to the functions, and called such functions *analytic*. This mode of representation is adapted to computation by successive approximations.

Riemann on the other hand may be said to have *geometrized* or *topologized* the analysis of functions of a complex variable. By a most ingenious model of connected sheets or membranes superimposed on a plane, for instance, he gave an intuitive picture of the properties of certain highly important complex functions, particularly those which take several different values for a given value of the variable. This development contributed greatly to *analysis situs*, or (as it is now usually called) *topology*, the geometry which studies the properties of surfaces, volumes, and the like which are invariant under a continuous group of transformations. We looked at this in [8.5–8.8].

❖ *16.6* ❖ *Conformal Mapping*

Only one detail, but an important and useful one, from among a multitude can be noticed here. The analytic functions [16.5] of a complex variable are of the most frequent occurrence in scientific and engineering applications. I must refer here to the discussion of partial derivatives [15.3]. Write z for the complex variable $x + iy$, $(i = \sqrt{-1})$, where x,y are real numbers. If the function $f(z)$ is separated into its real and imaginary parts, say $f(z) = u + iv$, where u,v are differentiable real functions of x,y, it can be proved that $f(z)$ is an analytic function of z in the sense of Weierstrass [16.5] if u,v satisfy the *Cauchy-Riemann equations*

$$\frac{\partial u}{\partial x} = \frac{\partial v}{\partial y}, \frac{\partial u}{\partial y} = -\frac{\partial v}{\partial x}.$$

For example, if $f(z) = z^2 = (x + iy)^2$, then

$$u = x^2 - y^2, v = 2xy,$$

and

$$\frac{\partial u}{\partial x} = 2x, \frac{\partial v}{\partial y} = 2x, \frac{\partial u}{\partial y} = -2y, \frac{\partial v}{\partial x} = 2y,$$

so z^2 is an analytic function of z. This of course is not meant to be evident.

Differentiating the above equations, the first with respect to x, the second with respect to y, we get

$$\frac{\partial^2 u}{\partial x^2} = \frac{\partial v}{\partial x \, \partial y}, \frac{\partial^2 u}{\partial y^2} = -\frac{\partial v}{\partial y \, \partial x}.$$

But under our assumptions

$$\frac{\partial v}{\partial x \, \partial y} = \frac{\partial v}{\partial y \, \partial x};$$

whence, by addition,

$$\frac{\partial^2 u}{\partial x^2} + \frac{\partial^2 u}{\partial y^2} = 0.$$

In the same way we derive

$$\frac{\partial^2 v}{\partial x^2} + \frac{\partial^2 v}{\partial y^2} = 0.$$

Thus each of u,v is a solution of Laplace's equation [15.5] in *two* dimensions. These simple facts, combined with a little geometry, are the basis for one of the principal uses of the theory of analytic functions, *conformal mapping*. If in a given plane two lines, straight or curved, intersect at a certain angle, the images of these lines in the conformal map will intersect at the same angle, although distances will usually be distorted. With this inadequate hint I must drop the subject.

❖ 16.7 ❖ Special Functions

The applications of mathematics have demanded the creation of a swarm of special functions, of which the simplest are the circular or trigonometric functions of school mathematics. Here again the field is too vast for more than a passing glance. The trigonometric functions will be discussed in the next chapter. But the idea underlying their characteristic property, *periodicity*, is so simple that it may be described in its first generality here.

Consider any periodic phenomenon—say the passage of the tip of the minute hand of a watch over the 12-o'clock mark. This passage is made at regular intervals of one hour. We say that the position of the tip is a *periodic function of the time*, with *period* one hour. Periodic phenomena permeate science. Wave motion is an instance. For this reason, if no other, periodic functions were extensively investigated by analysts ever since the time of Fourier and even earlier.

Expressing the periodicity in the above example algebraically, we write $f(t + 1) = f(t)$, which is read, 'function of $t + 1$ is equal to function of t.' Here 'function' may be interpreted as position expressed in terms of time t. The thing to be noticed is that the numerical value of $f(t)$ is unaltered when we replace the variable t by the linear expression $t + 1$, depending only upon t. Thus the *value* of the function is *invariant under a particular linear transformation of the variable*.

Why stop here? Poincaré in the 1880s went much farther, and considered functions invariant under *groups* (in the technical sense explained in Chapter 9) *of linear transformations of their variables*. The result was a new kingdom of analysis. As a by-product of all this, Poincaré *solved the general algebraic equation of the nth degree* by showing how its n roots can be expressed explicitly in terms of some of the functions he had created (compare [9.7]).

Finally, functions with any finite number of periods received much attention from the 1880s to the early 1900s. These alone supply enough for a life's work.

❖ *16.8* ❖ *Generalizations*

What is perhaps the most striking generalization originated in the work of V. Volterra (1860–1940) and his school in the 1880s and 1890s. In a word, Volterra investigated functions of a *non-denumerable infinity of variables*, a non-denumerable infinity [4.2] being as many as there are of points on a straight line. For example, instead of looking at a curve as a relation between the coordinates of any point on it, we may consider *the curve itself* as the variable thing, and see what happens as one curve shades into another. The curve however, viewed otherwise, is the set of all its points, and this set is non-denumerably [4.2] infinite.

This, and what grew out of it, appears to be the true mathematical approach to all those physical problems where all the past history of a given thing has to be taken account of in predicting the future. For example, a steel bar, when magnetized and then demagnetized, exhibits more or less permanent modifications which must be included in the mathematical analysis. Here and elsewhere, in economics for instance, the theory of *integral equations* and its modern extensions, developed largely since 1912, were promising leads. It should be noted, however, that this approach to the mathematics of economics was bypassed in the 1940s by proponents of the mathematical theory of games, invented by J. von Neumann (1903–), which proceeds on quite different lines. The subject of integral equations originated with Abel and R. Murphy.

Murphy (first half of the nineteenth century) was a clergy-man and went the same disastrous way as Hamilton. In passing, I believe that not sex but alcohol is the snare that mathematicians have to look out for.

Roughly the distinction between integral equations and those of the classical physics is this. In the classical mechanics and physics it is *rates of change* which enter the equations (*differential equations*); in the other work it is the *inverses* of such rates, or *integrations* (infinite summations) which appear. From a given relation between these it is required to disentangle the functions which are integrated. To complicate (and, paradoxically enough, simplify) matters further, in 1906 H. Lebesgue (1875–1941) revolutionized integration itself.

In the first flush of discovery some analysts in their enthusiasm predicted that integral equations and their generalizations would supplant the differential equations which had dominated physical science for over two centuries. A reason for the prediction was that solving an integral equation is equivalent to solving a boundary-value problem [14.8]. The difficulties of producing usable solutions of integral equations quenched the premature enthusiasm. Nevertheless integral equations are part of the standard equipment of the mathematical physicists, and are regularly taught today in technical schools of the better grade.

Chapter 17

WAVES AND VIBRATIONS

❖ 17.1 ❖ Periodicity

Waves and vibrations play a predominant part in the physical sciences and their applications, from the electromagnetic theory of light to the calculation of the oscillations of tall buildings during an earthquake.

Until the advent of special relativity in 1905, many periodic phenomena, such as those observed in the study of light, were pictured as actual waves in an existent ether. With the fading of the ether as a physical reality, it quickly became unfashionable to think of light as a vibration in a material medium. But it may still be quite helpful for one type of mind to continue visualizing the phenomena of light in this outmoded way. Although mechanical models passed out of favor with the modern quantum theory (1925), thinking in terms of them is inherently no less respectable than translating everything into mathematical abstractions. Both methods yield purely symbolic representations of nature, and it does not follow that one is more 'realistic' than the other. It is all a question of who is doing the thinking.

For example, Faraday was no mathematician, nor was Hamilton much of a physicist. Yet Faraday made fundamental discoveries in electricity and magnetism by a process of reasoning which, although not cloaked in the orthodox symbols, was essentially mathematical with its lines of force and the like. Such at least was the opinion of two great mathematical physicists, Maxwell and Lord Rayleigh

(1842–1919). Hamilton, on the other side, scarcely glancing at physical imagery, discovered conical refraction [16.4] by mathematical analysis.

A 'wave' to a mathematician is usually nothing more than a convenient term for describing the periodic character of certain functions obtained as solutions of differential equations. If a periodic function be plotted as a graph, the pattern of the graph falls into repetitions of one subpattern,

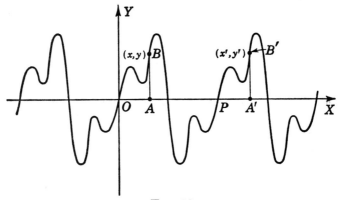

Fɪɢ. 30

such as the part from O to P (see Figure 30). Here the periodicity is evident to the eye. The entire curve might represent approximately the cross section of a choppy sea, in other words a superposition of several 'waves.' But this is not the sense in which 'wave' is used in mathematics. That sense is the following, easily seen on the diagram.

The length OP measures the *period* of the graph, that is, of the function plotted. Take any point A on the axis and erect the perpendicular AB cutting the curve in B. Do the like at A' where $AA' = OP$. Then $A'B' = AB$. That is, at points one period-length apart, the perpendiculars are equal in length and have the same sign. Starting at A' we repeat the process and reach the same conclusion, and so on all along the curve.

To restate this in symbols, let $y = f(x)$ be the equation of the curve, and p the length of the period OP, measured in the same units as those used for the distances along OX. Then if (x, y) are the coordinates of the point B on the curve, $OA = x$, $AB = y$, so that $AB = f(x)$. Also

$$OA' = OA + OP = x + p = x',$$

$A'B' = y' = y$ and $y' = f(x')$, since (x', y') is on the curve. Thus we have $y' = f(x + p)$, $y = f(x)$, $y = y'$, so that

$$f(x + p) = f(x).$$

In the same way

$$f(x + 2p) = f(x + p);$$

thus

$$f(x + 2p) = f(x).$$

Generally, if n is any *integer*, positive, zero, or negative,

$$f(x + np) = f(x).$$

If n is negative we have gone to the left of O. All this is summed up by saying that $f(x)$ is a *periodic function of x with the fundamental period p*. The *fundamental period* is the *least* number for which the curve 'repeats' as described. Hereafter I shall say simply 'period.' It is the concept of *periodicity*, or of a *periodic function*, which is important, not its graphical representation by a wavy curve.

To see that a graph which looks 'wavy' does not necessarily represent a wave in the sense of a wave in water or in any other medium, consider a cardiogram. This is an electrically recorded graph of the heart beats, and in records of excitement or severe disorders it looks like the cross section of a very choppy sea indeed. Yet the heart beats have nothing to do with 'waves' in the ordinary sense.

Actually the cardiogram represents the periodic fluctuations of the energy of the heart beats.

Periodic phenomena in our daily lives are so common that we seldom notice them. In extreme cases, such as that of the English gentleman who cut his throat one morning before breakfast because he could not endure the sudden knowledge that shaving is a periodic function of the time, it takes a disaster to make us aware of the monotonous reiterations of our lives. At approximately equally spaced intervals of time we do the same things over and over again in approximately the same way. Our very breathing and heart beats are periodic. Even death is no escape. We shall rise again, we are assured, and the Hindus must submit to a yet more distressing periodicity.

Nature too is a slave to periodicity. The seasons, the positions of the planets, the tides, darkness and light, sunshine and moonlight, all these and scores of others are periodic or approximately so. In some phenomena the periodicity is sharply definite and easily recognized. In others periodicity is perceived only after a long lapse of time, and then it appears only after several superimposed periodicities have been analyzed out. Such, for example, is the case with sunspots, and many industrious computers have struggled for years to detect periodicities in the weather and in the recurrence of earthquakes. Should even one earthquake period be found to a fair approximation for California or Japan, it is probable that much loss of life and property could be averted. As for the weather, something more than conversation at last promises to be done about it with a tremendous new calculating machine being built as this is written. According to the designers, the machine will tell us the weather accurately—for how long in advance I hesitate to say. We shall see.

The department of mathematics which detects periodicities and analyzes them—when possible—into simpler ones is, curiously enough, called *harmonic analysis*, from its origin in the theories of sound and vibrating strings. A musical note is compounded of its 'fundamental' and the successive 'harmonics' which give the note its distinctive 'quality.' The periods of the harmonics are submultiplies (rational fractions) of the period of the fundamental.

Another kind of wave phenomena, that of *damped vibrations*, is as common as the strictly periodic waves already described. The reader is no doubt familiar with one perfect

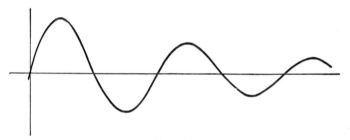

Fig. 31

instance of such vibrations in the Mouse's Tale as recorded in *Alice in Wonderland*, where the printed tale is presented in the sinuous curve of the mouse's tail. Lewis Carroll (1832–1878) perpetrated one pun intentionally and possibly two, for the vibrations of the tail would necessarily be somewhat damp after the mouse had swum the pool of tears.

A more commonplace example is that of a vibrating tuning fork. The resistance of the air and other causes quickly damp out the vibrations. If a bristle be attached to one prong of the fork and a smoked glass be drawn out evenly under the bristle, the tip will trace out a curve of damped vibrations (see Figure 31). The record of the heart-beats of a dying man would be similar but more complicated.

❖ *17.2* ❖ *The Alphabet of Periodicity*

Periodic phenomena are analyzed mathematically by what is known as *Fourier's theorem* after its discoverer. In addition to being one of the most astonishing results in the whole range of human knowledge, Fourier's theorem is mathematically indispensable in the theories of heat, light, sound, electricity, and elsewhere in physics. Fourier came upon the theorem while he was creating his mathematical theory of heat conduction. Not even attempting to indicate the refinements indispensable for an exact account, I shall describe in the succeeding section what the theorem states. It is necessary first to recall the definitions of the sine and cosine in trigonometry. This is no more difficult than reading the time on a clock.

In passing it may be remarked that trigonometry is often thought of as the mathematics of calculating sides or angles of triangles when sufficient data are given, and we might be inclined to think surveying the most important application of trigonometry. All this compared to the simple thing next explained is almost trivial for the current phase of civilization. It is because the sine and cosine are periodic functions of their variables that they are of importance in modern science and technology. These functions are the natural alphabet of all periodic change (see Figure 32).

As in Cartesian geometry [7.2], we lay down two mutually perpendicular axes intersecting at O, and we retain the conventions about distances to the right being positive, those to the left negative, those up positive, those down negative. Choose any convenient unit of length, and with a radius equal to this unit describe a circle with center at O. This is called the *unit circle*. Let this circle cut OX at A. Along OX lay off AB equal in length to the circumference of the unit circle. (Since the radius is 1, and the circumference of a circle is 2π times the radius, AB will be 2π units in length.) Let the point P start at A and trace out the circle in the direction of the arrow—'counter-clockwise.' Denote the angle AOP by a. As P proceeds from A to Q, to A', to Q', to A, to Q, . . . and so on, like the tip of a clock hand not stopping after a complete circuit, P sweeps out all angles from zero degrees to 90, to 180, to 270, to 360, to 450, . . . and so on. If P traces out the circle in the opposite direction ('clockwise') from A to Q', to A', to Q,

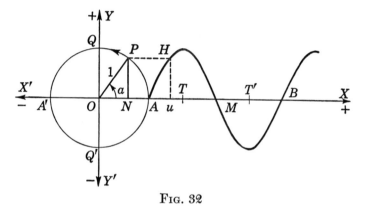

Fig. 32

to A to Q', . . . , we say, *by definition*, that the angles swept out are the *negatives* of those swept out before.

Let P be *any* point on the circumference of the circle. Drop PN perpendicular to AA'. The angle $NOP = AOP = a$ (a is the number of units of angular measure in the angle). Since AB is equal in length to the circumference of the circle, we can *represent* the angles from 0 to 360 degrees along AB as follows. To 180 degrees corresponds AM, where M is the mid-point of AB, since A' is halfway round the circle; to 90 degrees corresponds AT, where T is halfway between A and M; 270 degrees corresponds to AT' where T' is halfway between M and B; and to 360 degrees corresponds AB. As P continues to revolve in the same direction we get the distances measured from A corresponding to angles greater than 360 degrees.

Suppose that the angle a is one nth of 360 degrees. To a corresponds AU, where AU is one nth of AB.

Notice that nothing is being proved. All this is a matter of laying down a perfectly arbitrary but practically workable *graphical representation for the measures of angles.* Numerous other schemes might be devised; this is the one in use—because it is simple.

Refer now to the figure. Since the circle is the unit circle OP is 1 unit of length. The length of NP *is called the sine of the angle* a, and is written sin a; the length of ON *is called the cosine of the angle* a, and is written cos a.

If NP is *above* XOX', sin a is *positive;* if NP is *below* XOX', sin a is *negative;* if ON is to the *right* of YOY', cos a is *positive;* if ON is to the *left* of YOY', cos a is *negative.* These follow from the definitions of sin a, cos a and the conventions as to sign in Cartesian (analytic) geometry, which we assumed at the beginning.

We wish now to represent graphically the 'march' of the function sin a as a increases from 0 to 360 degrees. To do this we erect at the point U corresponding to the angle a the perpendicular UH equal in length to $NP(NP = \sin a)$, and draw this perpendicular *in the same sense* (above or below) as NP. Imagine this done for all the angles from 0 to 360 degrees. Then the tops of the perpendiculars will lie on a continuous curve, represented in the figure, one complete period of *the sine curve* (from A to B).

The cosine curve is graphed similarly. We shall leave it to the reader to see that the cosine curve is of exactly the same shape as the sine curve, and is in fact obtainable from the complete sine curve by shifting the latter through a distance AT (one-quarter of AB) to the left. The whole sine curve is the repetition of the part for one complete period (shown from A to B in the figure) indefinitely in both directions. For, as we see, angles greater than 360 degrees merely repeat the sines over again of those up to 360. Further, *negative* angles (as already defined) will give the same curve continued to the *left*.

Finally, then, the complete sine curve has as its graph the repetition, indefinitely in both directions, of the part from A to B.

The likeness of this curve to the profile of a train of waves on water is evident. But once again we must emphasize that this is *not* a profile of *any* material wave; it is a purely schematic representation of a *periodic variability,* as in that of sin a, *when a varies continuously.* The *values* of the *function* recur periodically as the variable (here a) increases or decreases continuously.

To complete what has been said in the definition of sines and cosines, we must add a word about the unit in terms of which we measure angles. Degrees have been mentioned, but the degree is an artificial unit of angular measure in no sense germane to the 'circular' description of angles. Possibly we use degrees because the earliest shepherd-astronomers of Sumeria rated the length of the year as 360 days and passed on their crude approximation to the pioneer mathematicians and astronomers of Babylon. It is known that the 360 came from the Babylonians; where they got it is not known. Wherever it came from, 360 is a primitive monstrosity undeserving of mathematical survival. It should have been pitched to the undiscriminating dodo or into the backwash of Noah's Babylonian ark when the decimal system of numeration was invented, if any artificial system was to be retained.

The natural unit of angular measure is the radian. One radian is the angle subtended at the center of a circle (*any* circle) by an *arc* whose *length* is equal to the length of the *radius* of the circle. Referring back to the diagram for sines, we see that AB represents 2π radians, where π is as usual the ratio of the circumference of any circle to its diameter.

Measuring the angle a in radians, we sum up everything the discussion has given us in the equations

$$\sin\,(a + 2\pi) = \sin a, \cos\,(a + 2\pi) = \cos a.$$

These say that *the sine and cosine are periodic functions,* their *period* being 2π.

One final remark applies to the following section. The graph of $y = \sin x$, as we have seen, is the wavy sine curve. The graph of $y = 2 \sin x$ will be obtained from that of $y = \sin x$ by stretching every y to twice its length. If, in Figure 33, (1) is $y = \sin x$, (2) will be $y = 2 \sin x$. To plot $y = \sin 2x$, note that $2x$ runs through its range of values twice as fast as x. Thus (3) represents $y = \sin 2x$. Similar remarks enable us to sketch $y = a \sin bx$, where a,b are any real numbers, directly from the graph of $y = \sin x$, and likewise for $y = a \cos bx$, $y = \cos x$.

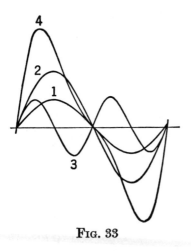

FIG. 33

Suppose now that we wish to 'add' any number of such curves. This is done by adding the y's for the same x on all the curves as x varies through the least range which includes a full period of each of the curves. In this addition we must attend to the signs of the addends, as in algebra, remembering that y above the axis is positive and y below, negative. The curve (4) in the figure is the sum of (2) and (3); its equation is

$$y = 2 \sin x + \sin 2x.$$

Conversely, if we had to plot this equation, we should first plot the curves $y = 2 \sin x$, $y = \sin 2x$, and then add them graphically. The procedure is the same for any sum of sines or cosines, or both.

❖ *17.3* ❖ *Fourier's Theorem*

Within certain limitations Fourier's theorem states that *any* periodic graph can be made up by adding a sufficient number of graphs of the forms

$$y = a_1 \sin x, \, y = a_2 \sin 2x, \, y = a_3 \sin 3x, \, \ldots ,$$
$$y = b_0, \, y = b_1 \cos x, \, y = b_2 \cos 2x, \, y = b_3 \cos 3x, \, \ldots ,$$

in which $a_1, a_2, a_3, \ldots , b_0, b_1, b_2, b_3, \ldots$ are numbers which depend only upon the particular periodic graph considered and which can be calculated when the equation of the graph is given or, more generally, when the function represented by the graph is defined. For example (Figure 34) the broken graph (continued indefinitely) is described by saying that it is periodic, with period 1, and that the function graphed is equal to 1 for all values of its variable greater than zero and equal to, or less than $\frac{1}{2}$, and is equal to -1 for all values of its variable greater than $\frac{1}{2}$ and equal to, or less than 1, with a similar description for the part to the left of O. Thus if $y = f(x)$ is the equation of this broken graph, $f(x) = 1$ for the first half of the complete period, and $f(x) = -1$ for the second half.

Note that at $x = \frac{1}{2}$ there is a sudden jump from $+1$ to -1 in the value of the function. Such a jump through a non-zero distance is called a *discontinuity* of the function, and one of the limitations mentioned at the beginning is that the graph shall have only a finite number of discontinuities in any finite interval.

The number of elementary components (simple sine or cosine curves) into which the periodic graph is resolved, or from which it can be compounded by additions, is not necessarily finite. But in any given instance a sufficiently large finite number of components will give the resultant curve to any required degree of accuracy.

Physically the theorem amounts to this: *any* periodic disturbance can be resolved into a sum of simple harmonic disturbances—the kind represented by the wavy sine and

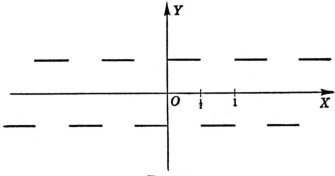

Fig. 34

cosine curves. It seems incredible that the broken-line graph above, for instance, could be so resolved into continuous wavy curves. But when we think that the periods of the wavy curves $y = a_1 \sin x$, $y = a_2 \sin 2x$, . . . get shorter and shorter as we go along the sequence, and that a_1, a_2, . . . may similarly decrease, the theorem is less miraculous.

Fourier's theorem is one of the most penetrating instruments of applied mathematics. In applications one component is often much larger and more important physically than the others; this corresponds to the 'fundamental' of the periodic disturbances producing the combined effect. To a first approximation only this component need be considered; further components added to the fundamental give closer approximations. Knowing the fundamental we can predict within ascertainable limits of accuracy the recurrence of the phenomenon under investigation—say the periodicity of sunspots. Among the earliest efficient calculating machines were the 'harmonic analysers' designed to break down a periodic function into its simple components. A limitation of these analysers was the practical difficulty of cutting accurate small gears.

Boundary-value problems [14.8] of certain very general types are solved by means of expanding given functions into Fourier series, that is, resolving the graphs of the functions as explained above.

Finally, Fourier's theorem is but an episode, although a capital cone, in the vaster theory of what are called *orthogonal functions*. These entered mathematics through science, and offer one of the comparatively rare instances where a mathematical theory demanded directly by scientific applications fitted naturally into the warp and woof of analysis as woven by pure mathematicians. It is usually the other way about: either the mathematics necessary for use in applications was already developed when the need for it arose, or the new demands necessitated the creation of novel analysis, often uncouth and not hewn into decent mathematical shape for decades.

✤ 17.4 ✤ *From Particles to Fields*

We have already described [12.1] Maxwell's mathematical prediction of wireless waves and its experimental vertification by Hertz in 1888. From that prediction and verification, as was noted, the whole wireless and radio industry developed, beginning commercially with G. Marconi's (1874–1937) sending of wireless signals across the English Channel in 1899.

Of course the waves *might* have been discovered if Maxwell had not predicted them from his mathematical theory of electricity and magnetism (and their connection with light), but they *were not* so discovered. We shall now amplify the earlier sketch [12.1] and outline the mathematics of Maxwell's prediction.

Maxwell started from the experimental researches of Faraday, translating the 'lines of force' and other physical imagery of that great unmathematical genius into the language of mathematics. Faraday is given full and ungrudging credit by Maxwell for his share. Indeed, to some, it has seemed that Maxwell, a modest man by nature, rather overdid modesty on this occasion. The ability to translate physical imagery into significant mathematical symbolism seems to be the rarest of all scientific gifts. Newton and Maxwell (sometimes called the Newton of the nineteenth century) had it in superlative degree; Faraday was entirely devoid of it, in spite of what Maxwell himself says of Faraday's essentially mathematical cast of thought. For the actual translation demands high mathematical skill of a purely technical kind, such as professional pure mathematicians have by instinct sharpened by training. There is not the slightest evidence that Faraday either possessed this talent, latent for lack of technical training, or was capable of acquiring it. Maxwell on the other hand was a born mathematician and might, had he chosen, have been one of the great pure mathematicians of the nineteenth century. In addition he was a born topologist and had extraordinary insight into physical phenomena. It was the same combination that Newton had, and Maxwell's scientific weight in the nineteenth century—and after—is comparable to Newton's in the eighteenth.

What now appears to have been Maxwell's main contribution to the theory of electricity and magnetism was his mathematical theory of the *field*. This is summed up in his famous equations, a special case of which I shall state presently. As Maxwell says,

> In electrical investigations we may use formulae in which the quantities involved are the distances of certain bodies, and the electrifications or currents in those bodies, or we may use formulae which involve other quantities, each of which is continuous through all space.
>
> The mathematical process employed in the first method is integration [15.6] along lines, over surfaces, and throughout finite spaces [volumes]; those employed in the second method are partial differential equations [15.3–15.5] and integrations throughout all space.

Maxwell then goes on to say that Faraday's method is essentially the second:

> He [Faraday] never considers bodies as existing with nothing between them but distance, and acting on one another according to some function of that distance [as in Newton's law of universal gravitation]. He conceives all space as a field of force, the lines of force being in general curved, and those due to any body extending from it on all sides, their directions being modified by the presence of other bodies.

Thus Faraday and Maxwell abandoned the Newtonian concept of 'action at a distance' and, in doing so, invited the mechanically minded to invent the all-space-filling 'ether' as the locus and everlasting habitation of all 'fields of force.' The etherists in their turn abandoned their hard-won position to the relativists, who abolished the ether in favor of a curved space-time. Numerous experiments were devised to detect the assumed notion of the earth through the hypothetical ether. In each instance the result was negative. The most refined of all the experimental measurements were those by A. A. Michelson (1852–1931) and A. Morley in 1887. By a curious historical coincidence, 1887 was the year in which Ricci and Levi-Civita elaborated the tensor calculus [10.3] without which general relativity would have been impossible. But for relativity, the ether might still be with us as a fact though unobservable.

The equations which gave Maxwell his prediction of electromagnetic waves traveling with the velocity of light were constructed after a long chain of reasoning. The equations were a translation of observed facts—in the last analysis; hence their permanency. They were first published in 1864, twenty-four years before Hertz announced his experimental researches which were to verify Maxwell's prediction. We shall merely state the equations; the meaning of the symbol ∂ was explained in [15.3].

The notation is as follows: κ is the specific inductive capacity, μ the permeability, c the velocity of light in free space; E_x, E_y, E_z are the components of the electric intensity E in directions parallel to the axes of coordinates at the point (x, y, z); H_x, H_y, H_z are the components of the magnetic intensity H. The equations for a homogeneous isotropic medium containing no free charge are

$$\frac{\partial E_x}{\partial x} + \frac{\partial E_y}{\partial y} + \frac{\partial E_z}{\partial z} = 0, \qquad \frac{\partial H_x}{\partial x} + \frac{\partial H_y}{\partial y} + \frac{\partial H_z}{\partial z} = 0;$$

$$\frac{\partial H_z}{\partial y} - \frac{\partial H_y}{\partial z} = \frac{\kappa}{c} E_x, \qquad \frac{\partial E_z}{\partial y} - \frac{\partial E_y}{\partial z} = -\frac{\mu}{c} H_x,$$

$$\frac{\partial H_x}{\partial z} - \frac{\partial H_z}{\partial x} = \frac{\kappa}{c} E_y, \qquad \frac{\partial E_x}{\partial z} - \frac{\partial E_z}{\partial x} = -\frac{\mu}{c} H_y,$$

$$\frac{\partial H_y}{\partial x} - \frac{\partial H_x}{\partial y} = \frac{\kappa}{c} E_z, \qquad \frac{\partial E_y}{\partial x} - \frac{\partial E_x}{\partial y} = -\frac{\mu}{c} H_z.$$

On seeing these equations any intelligent mathematician would take the simple technical steps leading to the elimination of H or of E. Doing so, he would make the wholly unexpected discovery that each of H, E satisfies the *wave equation* of mathematical physics. In this equation c appears in the right position to tell what will be the velocity of the propagation of H and E. The wave equation itself expresses the fact that whatever satisfies it is propagated through space in the form of waves—a periodic disturbance in the sense already explained. The wave equations for E and H showed that electromagnetic disturbances are propagated in a vacuum as waves traveling with the velocity of light. The vibrations (waves) are transverse to the direction of ·propagation—like those of light.

From this, among other blessings, followed unlimited advertising at your own fireside. The next on the program will be robot aerial torpedoes directed by wireless, predicted as early as 1930. The torpedoes in one form or another are here. If they could talk they might parody General Pershing's carefully rehearsed *blague* when he landed with the American troops in France during World War I, "Lafayette, we are here!" substituting Maxwell for Lafayette.

Chapter 18

CHOICE AND CHANCE

❖ 18.1 ❖ *Probability*

The heading of this chapter is lifted from a book which has fooled many a prospective buyer in secondhand bookstores. Opening the small brown volume the normal customer, expecting to find an exciting tale of love and adventure, was shocked to discover page after page of algebraic formulas. Usually *Choice and chance* was hastily returned to the shelf, especially after a glance at the title page revealed the author as a clergyman. But occasionally some inordinately curious mortal would sample a page or two near the beginning, surrender himself to the fascinations of the tale, and finally (after paying fifteen cents for the book) walk out of the store with the germ of an exciting adventure tucked under his left arm.

The theory of *choice*, a subdivision of the vast domain of *combinatorial analysis*, answers such questions as "How many committees of 100 can be made up, each consisting of 30 Americans, 20 Englishmen, and 50 Eskimos, if there are 80 Americans, 90 Englishmen, and 2,000 Eskimos available to serve?" Here it is a question of choosing 30—20—50 from 80—90—2,000, and we wish to know in how many ways the choice can be made. The number of ways is so enormous that I shall not attempt to write it out.

Chance goes a step farther. Suppose there are precisely 10 of the 80 Americans who are bald, 7 of the 90 Englishmen who have black hair, and 1,023 of the 2,000 Eskimos who have red hair. What is the *probability*, or *chance*, that in a

374

committee chosen at random of 30 Americans, 20 Englishmen, and 50 Eskimos, precisely 9 of the Americans shall be bald, 1 of the Englishmen black-haired, and all the Eskimos red-haired?

The two kinds of problem supplement each other. Before the second can be solved we must know in how many ways it is possible to choose the designated bald-black-red committee from all the material available. Suppose there are F possible choices, and suppose the total number of ways of selecting a 30—20—50 committee (as in the first problem) is T. Then the required probability is F/T, the *ratio of the number* (F) *of favorable cases to the total number* (T) *of cases possible.*

As a simpler problem, what is the probability of throwing a 3 and a 2 in a single throw of two dice? The favorable cases—those we want—are exactly two in number (die A may be 2, when die B must be 3; or die A may be 3, when die B must be 2). But there are in all 6 $\times$ 6, or 36, ways in which the two dice may fall—since die A may come 1, 2, 3, 4, 5, or 6, and with whatever way die A falls, there are precisely 6 ways in which die B may fall. Thus the probability of a throw of a 3 and a 2 is 2 in 36, or $\frac{2}{36}$, $= \frac{1}{18}$. This means that in the long run—and sometimes a very long run indeed, as any gambler knows—the designated throw will occur (very approximately) once in 18 times. This is better expressed by saying that in a *large* number of throws with two dice, about one-eighteenth will be the throw 2, 3.

The last may be taken either as a definition or as the disheartening verdict of experience. Intuitively we sense what the probability $\frac{1}{18}$ means for the throw 2, 3, although we should be hard put to it to justify our intuition either by logic or by experience. It is in fact impossible to do so. Yet, following Laplace, one of the great developers and appliers of the mathematical theory of probability, we shall take as the *definition* of the *probability* of an event the fraction F/T, where F is the number of cases favorable to the event, and T is the total number of cases that are 'equally possible.' This, admittedly, is vague to the point of nonsense. It will not bear critical analysis. Still, as Laplace said (in effect), the mathematical theory of probability is only common sense translated into arithmetic; and if the definition is not common sense, what is it?

The definition can be made a little more reasonable as follows. The probability that a coin will fall heads in a single throw is $\frac{1}{2}$. We arrive at this by noting what happens in a *large* number of throws. Suppose we keep a record of a great many throws, grouping them in sets of 10:

$$10, \ 20, \ 30, \ 40, \ 50, \ 60, \ 70, \ \ldots$$
$$4, \ \ \ 9, \ 16, \ 22, \ 26, \ 29, \ 37, \ \ldots$$

where the number of times that heads fell is written under the total number of throws up to the stage indicated. We thus get a sequence of 'frequency fractions'

$$\tfrac{4}{10}, \ \tfrac{9}{20}, \ \tfrac{16}{30}, \ \tfrac{22}{40}, \ \tfrac{26}{50}, \ \tfrac{29}{60}, \ \tfrac{37}{70}, \ \ldots,$$

and on turning these into decimals we observe that they are all close to .5 and that they get closer to .5 as the sequence continues.

Actual experience shows that the greater the number of throws the more nearly are one-half of them heads. Accordingly we assume that in a sufficiently large number of throws the ratio of the number of heads to the total number of throws will approach as close as we please to $\frac{1}{2}$.

Similarly in all cases. If the sequence of 'frequency fractions' has a limiting value, we define that limit[1] to be the

---

[1] 'Limit' is used here only in the intuitive sense indicated. Strictly, 'limit,' as defined in the calculus, has no meaning here, although such 'limits' are frequently used, rather paradoxically, to *abridge the calculation of probabilities*.

probability of the event in question. But obviously we cannot say whether or not such a limit exists. Common sense, following Laplace, assumes that it does.

It should be remarked that this is not the only definition of probability. The logical or philosophical objections to this definition are fairly evident and need not be gone into here. Since about 1920 an attempt to evade these difficulties by a more sophisticated definition has made considerable progress. But any claim that the modern return to the a posteriori definition of probability is on the highroad to complete success is optimistic. Precisely the same objections appear in this reopened approach as those that make the foundation of mathematical analysis on an unobjectionable basis a task for our successors. One advantage of the 'frequency' definition—the one adopted here—is its appeal to intuition. We feel that we know what the definition means, whether or not much if anything of it remains when it is picked to pieces.

The problems of selecting the committee and the one on dice might suggest that the theory of choice and chance is little more than a mathematical curiosity. This is far from the fact. Indeed 'probability' has been called the most important concept in current science, "especially," as Russell remarked, "as nobody has the slightest idea what it means." That probability is of scientific importance will be suggested by the few examples of its numerous uses to be described presently. For the moment it will be interesting to glance at the origin of the theory.

The real beginning of the theory of probability goes back to 1654, when Pascal and Fermat laid down the fundamental principles in a short correspondence. Most mathematical disciplines have had respectable enough parents; the progenitors (not Fermat and Pascal) of the theory of probability were thoroughly disreputable. For that reason, perhaps, the theory is all the more precious to us. Two gamblers had fallen into a dispute over the division of the stakes in an unfinished game. If a certain number of points were required to win the game, and the score of each player at the time of quitting was given, how should the stakes be apportioned between them? Clearly this comes down to calculating each player's probability of winning. The problem was too much for the baffled gamesters. They put it up to the pious Pascal, who shared it with the practical Fermat. In a few weeks these two mathematicians had laid down the fundamentals of the theory of probability. So every time you pay a premium on your life insurance you may breathe a prayer for the repose of the souls of those two gamblers— unless you grudge the premium, when you may curse the mathematicians. What you pay is based on the probability that you will die within the year. You bet that you will, the company bets that you won't. If you win, you lose—a modern version of the old saying that he who would save his life must lose it.

Pascal went even farther in his famous wager with the devil. The expectancy of winning a prize in a lottery is the value of the prize, say in dollars, multiplied by the probability of winning. Pascal argued that even if the average sinner's probability of winning eternal bliss may be very small indeed, yet the value of the prize is so great that it will pay him to take a chance, reform his ways, and live a godly, righteous, and sober life.

Although probability belongs to the discrete side of mathematics, appeal must be made to the continuous for the actual performance of the necessary calculations. This is not a theoretical necessity but a practical convenience. For example, the following type of problem appears repeatedly in statistical work, including statistical mechanics.

In how many different ways can b boxes be filled with t things, one in each box, from a collection containing exactly t things, all different? We shall assume that t is equal to, or greater than, b. Elementary reasoning gives the answer at once.

Imagine the b boxes arranged in a row. Any one of the t things may be put into the first box; this leaves any one of $t - 1$ for the second, then any one of $t - 2$ for the third, and so on, until $t - b + 1$ are left for the bth box. But if there are M ways of doing one thing, and N of doing another, there are $M \times N$ ways of doing both together. By repeated application of this we get

$$t \times (t - 1) \times (t - 2) \times \cdots \times (t - b + 1)$$

as the required number of ways of filling the boxes.

In practical examples, as in statistical mechanics, t may be a very large number and b a large fraction of t. If t is only a million and b half a million, we have to multiply together all the numbers 1,000,000, 999,999, . . . , 500,001. The whole human race toiling eight hours a day would probably take several years to do the multiplication, and when it was done the resulting number would be useless for practical purposes. What is wanted for any t,b is an approximate formula in terms of t,b which will be usable. Moreover this formula should contain a term estimating the amount by which the approximation is in error. The discovery of such formulas is done by analysis, so once more we are thrown back on the calculus and its modern developments.

❖ 18.2 ❖ *Pies, Flies, and Concrete*

To many sufferers 'statistics' means diagrams of pies and dollars cut up to represent the amount of the national wealth expended annually on pastry, cigarettes, cosmetics, and so-called 'defense.' This is one kind of statistics, but it does not tell us very much. Mathematical statistics, based on the theory of probability, goes far deeper, analyzes complicated masses of data, interprets them, and in some instances makes predictions. Thus we might ask what plausible or probable inferences may be drawn from the detailed analysis of a *random sample* from a very large number of specimens. We are interested in conclusions about the whole of the 'population' from which the random sample is taken. If the 'population' were human we might investigate a random sample—say 10,000—in order to find out whether thin persons were more likely than fat ones to be gluttons.

Here again we run into the same sort of difficulties that we meet in trying to understand probability. How can we be sure that our sample really is 'random' and representative of the entire population? These doubts need not trouble statisticians, as their findings are usually tested by experience.

An amusing instance from American history will illustrate the point. Random sampling of the voters made Mr. Dewey the winner and Mr. Truman the loser in the presidential race of 1948. It may have been that the sampling was biased, neglecting forgotten men and women who really work, and work hard, for their meager livings.

Then again there is the matter of extrasensory perception, where an uncritical use of probability and sampling produced astounding conclusions, such as that of the intelligent horse that received messages at a distance of 250 miles. If not extrasensory, this was at least extra-common-sensory, and therefore possibly a doorway into the Great Unknown.

How random is 'random'? And how impartial are statisticians? When something new is indicated, direct appeal to the facts settles the matter, but not always readily.

For example, a certain statistical investigation, full of high-powered mathematics, yielded the unexpected conclusion that vaccination against smallpox is a waste of time and money. This, of course, is the sort of applied mathematics that always delights one large section of the community, no matter how much it may annoy the medical profession. For my own part I have swallowed mathematics enough in my life to be immune to just one more dose, and I shall continue to get smallpox vaccinations whenever I contemplate a vacation in any of the filthier parts of the North American continent.

It may be a prejudice on my part, but I believe that many of the more loudmouthed leaders in education and a majority of stupidly led teachers should take out similar insurance against the less desirable consequences of indiscriminate intelligence testing. They might be less enthusiastic were they to swallow a liberal dose of the higher mathematics, including the theory of probability, on which the useful art of intelligence testing is based. The same holds for most applications of mathematics to the actual world. Only a decently critical familiarity with *all* the assumptions underlying a particular mathematical formula can teach us what not to take too seriously when the formula presents us with an impressive-looking number. Mathematicians are not, as a rule, credulous; their clients almost invariably are.

An interesting application of statistical methods to the biological sciences may be mentioned in passing. In the study of heredity the ultramicroscopic genes—the hypothetical units or carriers of heritable characteristics—have been assigned their relative positions in the chromosomes by what amount to statistical methods. By classifying the offspring of numerous generations of fruit flies, and recording the frequencies with which various characteristics—red-eye, sterility, mosaic hairiness, and so on—occur, it has been possible to draw up maps of the way in which the genes are distributed, and these maps reveal much that makes the flies what they cannot help being. The maps are purely schematic representations with the inherent power of prediction within reasonable limits, analogous to those of organic compounds in chemistry. The predicted genes have been observed by modern electron-optical technique.

If such experiments were conducted on human beings there would no doubt be a hullabaloo surpassing that which greeted the Copernican theory when it ousted man from the center of the universe. But for the present all is serene. The flies pass no resolutions deploring their merry pastime, and human beings, in spite of their manifest lechery, breed far too slowly to be suitable laboratory material at this stage of the libidinous game. However, it is not unlikely that our descendants a century or two hence will exhibit in their eugenic persons the mathematical theory of statistics as interpreted by some luetic dictator or crackpot savior of the race. Chance will then be eliminated from the human scene because no man will have choice. Man, to say nothing of woman, will mate as he is commanded to mate—and will probably enjoy it. Mussolini and Hitler, those vociferous proponents of wholesale human breeding, may be reincarnated. Of course, 'defense' in the shape of an atomic war may sterilize the race, or transmute its genes to those of subhuman monstrosities incapable of carrying food to their mouths. This might be the happy issue out of all its afflictions that our race has been praying for all these centuries. Some military propagandists assure us that this will never happen; some civilian geneticists are positive that it will. Take your choice while you have a chance left.

The practical ramifications of the statistical method are

everywhere in technology and the social sciences. In geodetic surveying, for instance, and in all extensive testing of machinery or mass-manufactured products, the measurements are reduced by the theory of errors of observation to isolate the most probable value of the thing measured. In the construction of dams, random samples of the rock on which the dam is to be built are taken—or not taken, as was the case in the 1930s in one California community with disaster as the price of cocksure political practicality. One such oversight cost a certain county about six million dollars. The concrete for the dam is also tested by random sampling.

The technical analysis of any large collection of data is a task for a highly trained and expensive man who knows the mathematical theory of statistics inside out. Otherwise the outcome is likely to be a collection of drawings—quartered pies, cute little battleships, and tapering rows of sturdy soldiers in diversified uniforms—interesting enough in a colored Sunday supplement, but hardly the sort of thing from which to draw reliable inferences. Either that or another six million dollars' worth of taxpayers' money gone down the chute.

❖ 18.3 ❖ *Statistics and Mechanics*

I have alluded [12.2] to a mathematical attack on the stability of our galaxy. The appropriate mathematics for such an enterprise belongs to what is called *statistical mechanics*. More than a passing glance at this intricate subject is out of the question here, but as it is one of the major fields of physics to which the mathematical theory of probability has been applied, I shall describe one of its details.

Ordinary mechanics reduces the investigation of dynamical systems to the solution of differential equations satisfy-

ing given initial conditions [14.8]. The grand object there is
to reduce the description of nature to central forces (as in
Newton's universal gravitation, for instance—although not
necessarily according to the inverse square law [13.3]) acting
between all pairs of material particles in the universe. It is
known that this program is not sufficiently inclusive. On
the purely technical side it raises prodigious difficulties.

A sort of compromise for complicated systems consisting
of swarms of mass-particles was worked out in statistical
mechanics, particularly by J. W. Gibbs (1839–1903), in
which the laws of ordinary mechanics are assumed to hold
for individual particles. More generally, the theory applies
to any system consisting of a large number of members each
of which can move—for example, a solid composed of
molecules, or a finite volume of gas. To obviate the difficulty of having to solve a 'practically infinite' number of
differential equations [15.4] for one system, a whole set of
systems is investigated statistically. This may seem like
throwing all the fat into the fire at once, but it luckily
turns out otherwise.

For simplicity consider first a single particle of mass m
moving along a curve. Its position $\bar{P}$ at time t on the curve
could be specified by the length of the arc from $\bar{P}_0$ to $\bar{P}$,
where $\bar{P}_0$ was the position at time $t = 0$. This length, say
q, is a *coordinate* giving the *position* of the particle; q is
assumed to be a function of t. The derivative of q with
respect to t is the velocity of the particle at time t [15.1];
this derivative multiplied by m is the *momentum*, say p.
Thus the motion of the particle is specified by $(q,\ p)$ in
which both q and p are functions of t (see Figure 35).

As t varies, q,p vary. We now plot $(q,\ p)$ as the coordinates
of a point as in plane analytic geometry [7.2], the values of
q being measured along the axis QOQ', those of p along
POP', for the successive values of t. As t varies, $(p,\ q)$ traces

out a curve. In an apparent way this *static* picture—the curve—represents the *motion* of the particle.

The set of all values of (q, p) is a two-dimensional manifold [8.3]. A curve in this manifold is a one-dimensional manifold which represents the motion of a particle.

All this can be greatly generalized. I shall not attempt to describe the most general situation. It will be enough to

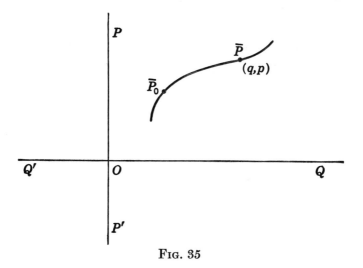

FIG. 35

give some idea of what is done by stepping up from one particle to two, to three, and so on, to n.

If instead of one particle we had n, we could represent the system in a $2n$-dimensional manifold [8.3], a typical point in this manifold having the coordinates $(q_1, \ldots, q_n, p_1, \ldots, p_n)$, where q_i, p_i are respectively the (q, p) for the ith particle. The manifold is technically called the *phase space* of the system. Thus the dynamical state of the system at a given instant is represented by a point in the phase space. The *phase* of the system is specified by the $2n$ coordinates of the representative point. As it has been assumed that the laws of Newtonian mechanics [13.3] hold for each constituent (particle, here) of the system, the representative point (q, p) of the system must traverse a path in the phase space consistent with these laws as the system changes in time from one configuration to another. From this it can be deduced that the direction of the curve in the phase space mapping the change of the system is determined for every point of the curve.

To proceed we fix our attention on the phase space, and as in discussing manifolds of many dimensions [8.3], imagine the situation for a plane (two dimensions) if we wish to visualize what is happening—the language of geometry takes care of the rest. Consider a small ($2n$-dimensional) volume, dV,[1] surrounding the representative point in the ($2n$-dimensional) phase space. We shall examine what is happening in dV under the following conditions.

We imagine a very large number N of independent systems each of which is identical with the first but with a different phase. We then have N points pigeonholed in different small volumes of the phase space. In other words we have the conception of a 'gas' or a 'fluid'—purely imaginary —whose 'molecules' are the representative points. It is therefore natural to analogize the 'flow' in a given region and we see what the 'density,' D, in the typical volume dV at the time t means. This density will be a function of the time t, and the coordinates of the typical point enclosed by dV; and the number of systems whose representative points lie in dV at the time t is DdV (compare with [15.5]).

We now assume that the distribution of representative points in the phase space is approximately continuous— indeed we have tacitly assumed this already in thinking of D as if it were the 'density' of a fluid. We can therefore add up all the DdV and take their limit as dV approaches zero.

---

[1] For the meaning of d in dV, see [15.1].

In short, we can integrate [14.7, 15.6]. The result of course will be only an approximation of the facts, as this kind of integration, strictly, is meaningless unless the distribution of points is actually continuous and not merely 'approximately' so.

The sum of all the $D\,dV$ is the total number of representative points. Hence [15.6] $\int D\,dV = N$, which may be written $\int \dfrac{D}{N}\,dV = 1$. But D/N is the probability, say W, that the representative point of a particular system selected at random shall lie in the volume dV. Thus $\int W\,dV = 1$. This may seem reasonable if we recall that 1 in probability represents certainty and that the probability of several *mutually exclusive* events is the *sum* of their probabilities; the representative point can lie in *only one* of the elementary volumes at a given instant.

The general equations of Newtonian dynamics and geometrical analogies suggest many interesting things to do. It can be proved, for example, that as a particular point moves about in the phase space the rate of change of phase-density in its immediate neighborhood is zero. Hence the representative points mimic the motion of an incompressible fluid (compare with [15.5]).

Much has been omitted from the preceding sketch of the preliminaries of the union of statistical methods with mechanics. From here on, conditions are imposed to connect the idealized abstraction with actual systems, such as those occurring in chemical reactions. The method is particularly useful in thermodynamics (the study of phenomena connected with heat), where the dependence upon probabilities is perhaps more evident.

Probability enters again in the present theory when it is used to show that a collection of a large number of particles having any initial distribution will tend to redistribute itself uniformly if left undisturbed. On reflection this is by no means evident. From this and similar arguments certain mathematical physicists reach the appalling—or comforting —conclusion of a universal 'heat death' in which all things will be at the same level of temperature—an infinite plenum of stagnant lukewarm glue.

Like many prophecies of nineteenth-century science, however, this one is not immune to modifications with advancing knowledge of the physical universe. In fact physicists of the 1920s found reasons for suspecting that the exceedingly queer integration we performed a little way back may not quite correspond to nature. We assumed that the representative points moved *continuously* in the phase space. Apparently this does not meet the demands of the quantum theory. If the motion is at least partly discontinuous—discrete—that integration and all its consequences would seem to be inadmissible.[1] But this is a matter for the physicists to decide. As already remarked [14.1] there is no evidence that incorrect mathematics necessarily leads to incorrect conclusions. Nor, of course, does correct mathematics necessarily lead to physically correct inferences. Frequently it does not, partly because of oversimplification in translating nature into mathematics.

An important divergence between dynamical theories and statistical is their split over the question of *reversibility*. It was noted [16.4] that Hamilton summed up classical dynamics in a 'stationary' principle. The mathematical expression of this principle involves the time t. The remarkable fact is that if t to be replaced by its negative, $-t$, the *initial* state of any system whose equations of motion are implied by the stationary principle is *completely* recovered

---

[1] Modern theories of integration, for example Lebesgue's [16.8], can take care of the *exact* physical situation, but such theories, as yet, fail to provide formulas which a practical physicist can *use*.

from the *final* state. Roughly, if the universe were an entirely dynamical affair, it might run as well backward as forward in time without violating the laws of dynamics. Such a reversibility does not occur in statistical theories.

An important instance of irreversibility occurs in thermodynamics where a consequence of the 'second law' is the heat death mentioned above. There seems but little to choose in the way of disagreeableness between a reversible universe and one which is irreversible. A suggestion—revived unwittingly from the ancient Babylonian myths—is that the universe may be periodic, so that everything will recur over and over again indefinitely, precisely as in the case of that dog, immortalized in Holy Writ, who returns to his vomit. A periodic universe would have the great advantage of combining all the undesirable features of the others and leaving nothing further to be anticipated with disgust. This theory of the 'eternal recurrence' was described by Plato and revived by Nietzsche among many others, who seem to have overlooked the logical detail that if the universe really is periodic, we must be unaware of the periodicity. But it has not yet been demonstrated, logically or empirically or theologically, that logic is either necessary or sufficient for the conduct of human affairs.

❖ 18.4 ❖ Is Probability Probable?

The apparently meaningless question above is not mine but Pascal's. After having brilliantly founded the mathematical theory of probability with Fermat, the devout Pascal turned aside for a moment to doubt whether what he had done had any meaning.

Pascal's question offers endless opportunities to philosophical debating societies. An interminable discussion might be carried on regarding the question itself. Is it a question at all? Or is it a pseudo question? It seems to suggest a paradox of the same kind as some of those which have worried mathematics since the 1890s and which have bothered philosophers of science, especially those who attend to the philosophical implications of probability in quantum physics.

Whether or not probability is probable, there is no doubt that it has become one of the ruling concepts of physical science. Our whole outlook on the meaning of such common-places of science as 'experiment' and 'observation' changed radically when probability in the 1920s drew a cataract of uncertainty across our eyes. We say that a physical 'ob-servable'—anything subject to laboratory measurement—does not have *a* value for a particular state of the thing observed, but *an average* value for that state.

All the applications of probability described earlier in this chapter are but little removed from technicalities of cal-culation. In none of them does probability play a part above that of the valet who makes his master's toilet less bore-some. Probability until the late 1920s was a means to an end. Classical statistics went so far because it was content to serve without criticism of itself or of its master. If ever a scion of mathematics was an obedient lackey of the sciences, it was the theory of probability.

But the spirit of revolution infected the humble servant along with his haughty master. No longer content to be a means to an end and a faithful servitor, probability pro-claimed itself both the end and the means to the end. Probability became the master of its master. And the crowning jest of it all was that this formerly despised lackey, born under a gaming table, set itself up as the dictator of fashionable science and its interpreter to be-wildered moralists begging from physical science a word to reassure them that man after all is the conscious director of his own will and not a mere straw blown hither and thither in winds of chance beyond his control. For in the eyes of distressed humanists there was a mote, not to say a beam, inserted there by quantum mechanics. They took their affliction with the panicky seriousness of a hypochondriac who at last gets something really the matter with him. They were at the point of death, or thought they were—in their case probably the same thing, for they demonstrated their ability to think themselves into or out of anything.

All this distress was occasioned by what quickly became a commonplace in theoretical physics, the so-called 'uncertainty principle' enunciated in 1927 by Heisenberg, with Dirac a founder of the modern quantum theory. In the older physics the possibility of measuring both the momentum and the position of an electron at any given instant was assumed as self-evident. But an analysis of the operations necessary for the dual measurement reveals an ineradicable indeterminacy: the more precisely either the position or the momentum is measured, the less precisely is it possible to measure the other. So the sharp certainty of prediction in the older mechanistic interpretation of nature is replaced in the new by a statistical fuzziness.

The larger the scale of the phenomena the less (usually) the margin of uncertainty. But this is no comfort at all to the confirmed seekers after eternal, indubitable truths. In phenomena at the atomic level of smallness 'almost anything' might happen, and we cannot predict with certainty from a given set of initial conditions exactly which one of several possibilities will actually occur. But man, in the flesh at least, is "a fortuitous concourse of atoms"—a 'system' in statistical mechanics. The fate of Faust was eternal blessedness itself compared to what confronted believers after Heisenberg upset them.

From here on the arguments for or against 'free will' based on physics are doubtless familiar to the reader. If not, they are readily accessible, so I shall gladly skip them. It suffices here to have suggested that the Servant of the sciences can rule when the Queen naps or becomes confused.

Chapter 19

"STORMING THE HEAVENS"

❖ *19.1* ❖ *Toward the Infinite*

The Servant now gives way to the Queen. While the Servant was somewhat uncritically exhibiting a small sample of her works, the Queen politely did not interrupt, waiting her chance to put her helper in her proper place, particularly about mathematical analysis and its infinities [4.2].

Infinity and the infinite have long had a singular fascination for human thought. Theology, philosophy, mathematics, and science have all at some stage of their development succumbed to the lure of the unending, the uncountable, the unbounded. "Only infinite mind can comprehend the infinite," according to one; "Cantor's doctrine of the mathematical infinite is the only genuine mathematics since the Greeks," according to another; while yet a third, contradicting both, declares that "the infinite is self-inconsistent, and Cantor's theory of the mathematical infinite is untenable."

Here we reach a frontier of knowledge, and progress since 1930 has halted. Some believe that mathematics may have to retrace many of the giant strides it made toward the infinite since G. Cantor in the 1890s practically completed his work; others foresee a steady progress in the direction already traveled. Some roads, however, are definitely blocked, as we shall see in the concluding chapter.

The simple fact seems to be that no one can say exactly where mathematics stands with regard to its supposed conquest of the infinite, and no one can sensibly predict its future. Equally competent authorities hold diametrically opposing views. But one thing seems to be admitted, the consistency of mathematical analysis has yet to be proved, if indeed it is provable at all on the Servant's level where it has been most useful.

With this caution against accepting anything in what follows as final, I may proceed to a short description of the kind of scaling ladders with which mathematicians, in Weyl's phrase, "stormed the heavens."

❖ *19.2* ❖ *How the Infinite Entered Mathematics*

The infinite entered mathematics early. Not to go too far back, let us glance at the problem of integration as it presented itself to Archimedes in the third century B.C. In calculating the area under a curve [15.6], appeal to the infinite occurs at the step of taking the limit of a sum of n rectangles of equal breadths as n becomes indefinitely great.

With the invention of the calculus in the seventeenth century and its applications to the finding of areas, surfaces, and volumes of all imaginable shapes, the performance of such infinite summations, or *integrations*, became a standard and unquestioned technique of the integral calculus [15.7]. Mathematical physics could not get on without integration. Consider for example the simple problem of calculating the work done as a *variable* force moves a body through a given distance, work being measured as force times distance in the proper units. On a higher level the variational principles already described [16.2, 16.3] involve integrations.

Even in the differential calculus [14.5] we meet the infinite at the very outset, and here we encounter a shattering disagreement between geometrical intuition and rigorous mathematics. It will be sufficient to state one geometrical application of differentiation. To draw a tangent line at a given point of a given curve necessitates the finding of the slope of the tangent line, and this is equivalent to performing a specific differentiation, or finding a specific derivative, in the technical sense [15.1] of the calculus. Now consider this. It is *intuitively evident* that we can always draw a tangent to a continuous curve at a given point of the curve. Intuition has deceived us; *there exist continuous curves which have no tangents at all.*

I admit gladly that this is shocking to common sense, for it shocked mathematicians when Weierstrass confronted them with such a curve in 1861. Weierstrass' curve is phrased analytically and is much more esoteric than others with the same tangentless peculiarity that can be described verbally and constructed (in imagination) by continued doodling of the kind inattentive schoolboys use to speed up the passage of boredom. Several interesting specimens of such curves are displayed in *Mathematics and the imagination* (1940) by Kasner and J. R. Newman.

If I may be autobiographical for a moment, I remember how the boys in school kept their minds alive during many a soporific session with Latin syntax or the Acts of the Apostles by surreptitious doodling of the kind that eventuates in tangentless curves and Peano's space-filling curves (1890), both continuous. Only one of the boys (it was not I) suspected that he might be doing serious mathematics. Unfortunately he was lost to the science for which he was born when he sacrificed himself to the severe competitive examination for the India Civil Service. He gained an honorable position and a fat salary. India in acquiring its independence from the British Empire dispensed with the I.C.S., while the unruly doodles of that boy's wasted hours with Bradley's *Arnold's Latin Prose* and the apostles continue to interest Indian mathematicians. The well-paid labor of that doodler's lifetime went down the drain, and he, poor devil, was forced to subsist on a greatly reduced pension. Like some others who take the wrong turning at a crucial juncture in their lives, he never suspected the cause of the stomach ulcers that made his life miserable until he was reduced to the comparative poverty which permitted him in his enforced leisure to resume the work for which he was suited. The moral of all this (I personally detest morals) for young mathematicians is, don't do what your teachers tell you is for your own good. Incidentally, this history reminds me of a placard in a shabby little café catering to the tourist trade: "Don't ask for information. If we knew anything, we shouldn't be here."

Now let us go back to summation a moment. The solutions of multitudes of mathematical and physical problems lead to unending sums. Here are three specimens where the individual terms in the sums are discrete. The infinite enters when n, in the sum of n terms of each of these series, becomes indefinitely great, or 'tends to infinity' through positive integer values.

$$1 - \tfrac{1}{2} + \tfrac{1}{3} - \tfrac{1}{4} + \cdots;$$
$$1 + x^2 + x^4 + x^6 + \cdots;$$
$$1 + \tfrac{1}{2} + \tfrac{1}{3} + \tfrac{1}{4} + \cdots.$$

These are almost pathologically simple, but they will do. The dots mean that the series are to continue *without end*, according to the law indicated in each case. Now, it can be proved that the first series *converges* to a definite, finite number as we proceed to infinity, adding and subtracting the fractions as they occur. If x is a real number [4.2], the second series converges only for such x as lie *between* -1 and $+1$; for all other real values of x the series *diverges;* that is, by adding a sufficient number of terms, the sum can be made to surpass any previously assigned number. The third series is divergent, although it does not look it.

It seems incredible that the sum of a sufficient number of terms of this series can be made bigger than a billion billion billion, but such is the fact. Another astonishing thing is that the first series is *not* equal to

$$(1 + \tfrac{1}{3} + \tfrac{1}{5} + \cdots) - (\tfrac{1}{2} + \tfrac{1}{4} + \tfrac{1}{6} + \cdots).$$

Each of the series in parentheses diverges, and it means nothing to subtract the second from the first.

Is it not clear that if a physical problem, say the calculation of a temperature, yields as answer a *divergent* series, then that answer has no physical meaning? Even hell cannot be infinitely hot. When such nonsense turns up we go back, revise our mathematics and reformulate the problem, or give it up.

One of the outstanding things Abel and Cauchy did in the early decades of the nineteenth century was to provide the first methods whereby the convergence of a series can be tested. Since then the theory of convergence has expanded like a thundercloud.

From the foregoing handful of examples we can appreciate the program of that great triumvirate Weierstrass, Dedekind, and Cantor, who in 1859 to 1897 undertook a thorough examination of the mathematical infinite itself. Another impulse to an attack on the infinite was the problem of irrationals. What does $\sqrt{2}$ mean, if it is *not the ratio of any pair of whole numbers?*

Dedekind's attack on irrationals is a modern reverberation of Eudoxus. If either falls under the counterattack of modern skeptics, both fall. Paradoxical as it may seem, the last conclusion is no novelty of the twentieth century. I. Barrow (1630–1677), a teacher of Newton and his predecessor at Cambridge, late in the seventeenth century acutely criticized Eudoxus. Barrow's objections to the logic of the great Greek have been repeated by the leading twentieth-century critics of the mathematical theory of the infinite elaborated by Weierstrass, Dedekind, and Cantor. If nobody listened to Barrow, the like cannot be said for Brouwer and his school.

Let us look at one or two of the central concepts of this controversial subject. Mathematical analysis [4.2]—the calculus [Chapter 15] and every luxuriant growth that has sprung from its fertile soil since the time of Newton and Leibniz—derives its meaning and its life from the mathematical infinite. Without a firm foundation in the infinite, mathematical analysis treads at every step on dangerous ground.

❖ 19.3 ❖ *Counting the Infinite*

Let us consider first what counting means. At a glance we see that the two sets of letters x,y,z and X,Y,Z contain the *same number* of letters, namely, three. We say that two classes contain the *same number* of things if the things in both classes can be placed in *one-one correspondence*, that is, if we can *pair off* the things in the two classes and have none left over in either. For example, we can pair x with X, y with Y, z with Z. We say that two classes are *similar* if the things in them can be paired in one-one correspondence.

Observe this simple fact: the classes x,y,z,w and X,Y,Z are *not* similar. Try as we will, we cannot find a mate for *some one* of x,y,z,w. The reason here is plain; the first class contains *four* things, the second only *three,* and four is greater than three. Everyone saw this for thousands and thousands of years, and for a wonder, everyone saw straight. The next took genius of a high order to perceive. G. Cantor was the modern hero of this.

Consider *all* the positive rational integers

$$1, 2, 3, 4, 5, 6, 7, 8, \ldots$$

and under each write its double, thus,

1, 2, 3, 4, 5, 6, 7, 8, . . .
2, 4, 6, 8, 10, 12, 14, 16, ...

How many numbers 2, 4, 6, . . . are there in the second row? Exactly as many as there are numbers altogether in the first, for we got the second row by doubling the numbers in the first. The class of *all* the natural numbers 1, 2, 3, 4, . . . is *similar to a part of itself*, namely, to the class of all the even numbers 2, 4, 6, 8, ... *There are just as many even numbers as there are whole numbers altogether.*

This illustrates a fundamental distinction between finite and infinite classes. *An infinite class is similar to a part of itself; a finite class is similar to no part of itself.* 'Part' there means *proper part*, namely, some but not all.

Actually, Cantor had been anticipated by Galileo in 1638. Galileo is the historical founder of the mathematical theory of the infinite. Because of its epochal significance, I transcribe the half-quizzical, half-satirical debate Galileo put into the mouths of two of his characters in his great dialogue on 'two new sciences.' It would be interesting to recall Galileo from the shades and ask him whether he really knew what he was about or whether he was just confusing a stupid Aristotelian in his own futile hairsplitting logic.[1] The sagacious Salviatus (*Salv.*) and the not-so-sagacious Simplicius (*Simp.*), whom the Inquisition suspected of being Galileo's caricature of the reigning Pope, discourse as follows.

[1] The extracts are from the first English translation, London, 1665, of Galileo Galilei, *Discorsi e dimonstrazione matematiche intorno a due nuove scienze*, Leida, 1638. The 1665 translation, *Galileus Galileus, Mathematical Discourses and Demonstrations*, and so forth, is much sharper than subsequent translations. Unfortunately, this edition is an excessive rarity, owing to the Great Fire of London, among other causes. It should be reproduced. Even the Library of Congress does not have a copy.

Salv.: . . . an Indivisible, added to another Indivisible, produceth not a thing divisible; for if that were so, it would follow, that even the Indivisibles were divisible . . . .

Simp.: Here already riseth a doubt, which I think unresolvable. . . . Now this assigning an Infinite bigger than an Infinite is, in my opinion, a conceit that can never by any means be apprehended.

To make the infinite plain even to Simplicius, Salviatus patiently explains what a square integer is before proceeding as follows.

Salv.: Farther questioning, if I ask how many are the Numbers Square, you can answer me truly, that they be as many, as are their propper roots; since every Square hath its Root, and every Root its Square, nor hath any Square more than one sole Root, or any Root more than one sole Square.

This is the kernel of the matter: the one-one correspondence between a part of an infinite class (here, that of all the natural numbers) and one of its subclasses (here, that of all the integer squares). Continuing the argument, Salviatus compels Simplicius to surrender.

Simp.: What is to be resolved on this occasion?

Salv.: I see no other decision that it may admit, but to say, that all Numbers are infinite; Squares are infinite; and that neither is the multitude of Squares less than all Numbers, nor this greater than that: and in conclusion, that the Attributes of Equality, Majority, and Minority have no place in Infinities, but only in terminate quantities . . . .

In modern terminology, two classes which can be placed in one-one correspondence are said, as we have noted, to be *equivalent* or *similar*. In Galileo's example, the class of all square integers is equivalent to the class of all positive integers.

I may quote here an interesting opinion of C. S. Slichter's (1864–1944) on what he, primarily an engineer, thought of

Galileo's incursion into the infinite. "What words do I put first among all the dictums of Galileo? I will surprise you by saying that my decision is upon the following"—the equivalence of the class of all square integers and the class of all positive integers. Is this so surprising when we consider what that dictum was ultimately responsible for in current mathematics and philosophy?

As another example, let us see that any two segments of a straight line contain the same ('uncountably infinite,' or 'non-denumerably infinite' [4.2]) number of points. (For brevity I am forced to omit many refinements which a mathematician would demand, but the following illustrates what is meant.) Suppose the segments AB and CD are of different lengths. Place them parallel as in the figure, and let *AC, BD* meet in *O*. Take any point, say *Q*, on

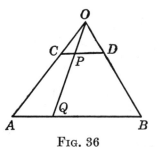

Fig. 36

AB, and join *OQ*. Let *OQ* cut *CD* in *P*. This sort of construction puts the class of all points on *AB* into one-one correspondence with the class of all points on *CD*.

Is there no escape? What about *postulating* that the points on a line are not dense everywhere, but strung like dewdrops on a spiderweb, and that any segment contains only a *finite* number of points? Such finite, *discrete* geometries were extensively investigated by American mathematicians in the early decades of the twentieth century by the postulational method. But to say that space—whatever scientists and others mean by space—is granular in structure and not continuous is too repugnant to habit to be acceptable. Nevertheless, in physics, energy parted lightly enough with some of its continuity in 1900 when M. Planck (1858–1947) quantized it, to avoid mathematical and physical absurdities. Instead of quantizing space, mathematicians at present prefer to overhaul their reasoning.

❖ *19.4* ❖ *What 'Is' a Number?*

An analysis [4.2] rests on numbers, I interpolate here an answer to the question of what a cardinal number is, say 2, or 3, or 4, or any other number which states 'how many.' The answer was given in 1884 by F. L. G. Frege (1848–1925), whose work passed almost unnoticed, possibly because much of it was written in an astounding symbolism which looked as complicated as a cross between a Babylonian cuneiform inscription and a Chinese classic in the original. It is the finest example of what happens when a mathematician does not write so that he who runs may read. Russell independently arrived at the same definition in 1901, and expressed it in plain English. Here it is: *The number of a class is the class of all those classes that are similar to it.*

This is not meant to be simple. It is profound, and it is worth pondering until one grasps intuitively whatever sense it may contain. Beside this gem of abstract thought the visions of the mystics seem material and gross.

It will be noticed that the word 'class' occurs essentially three times in the definition. In earlier chapters I repeatedly referred to classes or sets, and we agreed to take the notion of a class as intuitively evident. But intuition is not sharp enough for critical diggers at the foundations of mathematics, and these uncompromising critics ask to be given something more penetrating. So far they have not been satisfied, and the best they have been offered is still G. Cantor's attempted definition of 1895: "*By a class we understand any collection into a single whole of definite well-distinguished objects of our intuition or of our thought.*" It is sufficient here to say that the notion of a set (or class) is one of the unremoved obstacles in Cantor's theory of sets. The logicians and theologians of the Middle Ages stumbled over similar obstacles.

The phrase 'the class of all those classes' is the core of the above definition of the number of a class. How meaningful is it? In 1902 Russell propounded this teaser: "Is the class of all those classes which are not members of themselves a member of itself?" Either "Yes" or "No" leads to a contradiction, as the reader may think out for himself. Frege had relied on the intuitive logic (or verbalism) of classes in his almost lifelong attempt to put arithmetic on an unassailable foundation. The second volume of his masterpiece (1903) closes with this acknowledgement: "A scientist can hardly encounter anything more undesirable than to have the foundation collapse just as his work is completed. A letter from Mr. Bertrand Russell put me in this position just as this work was all but through the press." Russell's letter communicated the above paradox to the unhappy but candid Frege.

Another paradox of a different kind—which also appears disguised in some kinds of mathematics—is that of the barber. In a certain village a barber shaves all those, and only those, who do not shave themselves. Does the barber shave himself? Family arguments over this one might well end in divorce.

There are several more. One closer to arithmetic is named after its inventor, C. Burali-Forti (1861–1931) who noticed it in 1897. It is so simple when the ideas to which it refers are stated that it is strange—in retrospect—to see how it could have been overlooked for as long as it was. But Cantor himself did notice it, to his discomfort. It concerns Cantor's transfinite ordinals. In Cantor's theory of the infinite there is a radical distinction between cardinal and ordinal numbers. For finite numbers and classes the distinction is essentially trivial. We assign to a finite class a label 1, or 2, or 3, . . . , which characterizes the class irrespectively of the order in which its elements are arranged, and a particular label, the cardinal number of a given class, is also the label of any other class whose elements can be placed in one-one correspondence with those of the given class. When the elements of a finite class are counted in a given order, the first element being labeled 1, the next 2, and so on to the last, the final label is also the cardinal number of the class. But for infinite classes Cantor showed that this coincidence of cardinal and ordinal no longer holds. There seems (to some minds—not to my own) to be no difficulty in 'imagining' an infinite ordered class, for example that of 'all' the integers 1, 2, 3, ... Beyond all these lies the 'infinite number' ω (omega); beyond ω lies $\omega + 1$, then $\omega + 2$, and so on, until $\omega 2$ is reached; beyond all these lies ω^2, and beyond this $\omega^2 + 1$, and so on 'indefinitely.' All seems (on paper) to follow naturally enough after the first step. Imagine all the odd numbers written down. Then 2, which is not odd, is the next in order. Cantor's ω, $\omega + 1$, . . . were to give a means for well-ordering his transfinite numbers: a class is well-ordered if its elements are ordered and each has a unique successor. Cantor thought he had well-ordered the series of *all* ordinal numbers. Burali-Forti pointed out that the well-ordered series of all ordinal numbers defines a new ordinal number which is not one of the series. This was the disastrous paradox.

I have frequently used the concept of all the points on a straight-line segment 'corresponding to' the real numbers. These points cannot be counted off 1, 2, 3, . . . ; the 'infinity' which they define is non-denumerable [4.2], in contrast to the denumerable infinity defined by any class whose elements can be counted off 1, 2, 3, ... An outstanding problem is to prove or disprove that there is no class whose elements are less numerous than all the points on a line segment and more numerous than the integers 1, 2, 3, ... In other words, do these two infinities exhaust the possibilities? The problem, of course, can be stated more exactly, but this is a sufficient indication of what it is about.

❖ 19.5 ❖ Dedekind's Cut

How did Dedekind tame the irrationals? We postulated [4.2] that $\sqrt{2}$ can be represented by a point on the line of all real numbers, lying somewhere between 1 and 2, and that, by approximating more and more closely, we can narrow the interval in which the elusive number lies. But to trap it alone, and not get a whole brood of undesirables in the trap at the same time, requires supreme skill.

Dedekind provided this in his famous 'cuts,' which can be applied at any point of the line of reals. We need consider only that kind of cut which separates all the rational numbers into two classes of the following sort: each class contains at least one number: every number in the 'upper' class is greater than every number in the 'lower' class. Further, the numbers of the *upper* class have *no least number*, those of the *lower* class have *no greatest number*.

We can now imagine the 'upper' and the 'lower' classes laid down on the line of real numbers [4.2]. Owing to those provisos about no greatest and no least in the respective classes, the two classes will, as it were, strive to join one another. But they cannot, because any number in the upper is greater than every number in the lower. The place where they strive to join is the *cut*, and it defines *some irrational number*.

To locate $\sqrt{2}$ as a cut, we put into the *upper* class all those positive *rational* numbers [4.2] whose *squares* are *greater than 2*, and into the *lower* class all other rational numbers. A moment's visualization will reveal that the elusive $\sqrt{2}$ is definitely trapped between the two classes and is in the trap *alone*. The Greeks proved the irrationality by a contradiction. If $\sqrt{2}$ is rational, it is of the form a/b, where a,b are integers without a common factor exceeding 1. By squaring from $\sqrt{2} = a/b$, we get $2 = a^2/b^2$, $2b^2 = a^2$. So a must be even, say $a = 2c$. Then $b^2 = 2c^2$, so b must be even. That is, a,b have the common factor 2. Contradiction.

The Dedekind cut is at the root of modern mathematical analysis. Another root of that ever-fertile tree is the vast theory of assemblages which, roughly, discusses among other things the properties of curves, surfaces, and so on, as *sets* or *classes of points*. An outstanding problem in the theory of sets is this: can the elements in *any set whatever be well-ordered?* For example, consider once more (see [4.2]) all the points on a segment of a straight line. Between any two points of the line we can always find another point of the line. How then shall we individualize this uncountable infinity of points and call each by its name according to any conceivable system of nomenclature? We do not know. A very famous postulate, Zermelo's of 1904, practically assumes that any assemblage can be well-ordered, for his unaccepted proof rests on a doubtful postulate. The postulate asserts that if we are given any set of classes, each of which contains at least one thing, and no two of which have a thing in common, then 'there exists' a class which has just one thing in each of the classes of the set. Why should this be true, if it is, of an *infinite* set of classes? This assumption, like all of the notions sketched in this chapter, has been repeatedly challenged. It is much less innocent than it looks.

In the following and concluding chapter I shall indicate deeper difficulties.

Chapter 20

BEDROCK

❖ 20.1 ❖ *Mathematical Existence*

In nearly all that we have seen, two great themes of modern mathematics predominated: the postulational method [2.2], and the mathematical analysis [4.2] founded on the real number system. Underlying both was modern logic [5.2] as it has developed since Boole's *Laws of thought* of 1854, and particularly since the attempt in 1910–1914 of Whitehead and Russell to show that all mathematics is an exercise in symbolic logic.

Much of this work before 1930 was instigated by the patent need for putting a sound foundation under the enormous mass of nineteenth-century mathematics—both pure and applied. To the uninitiated it may seem a very queer proceeding to build up vast systems of knowledge without seeing first whether the foundations will bear the superstructure. Mathematics in the nineteenth century did precisely that. As weaknesses began to appear in the foundations, and one part or another of the colossal edifice crumbled, mathematicians made hasty repairs and went on building, until more serious faults made themselves evident, and so on well into the twentieth century. Who shall criticize the builders? Certainly not those who stood idly by without lifting a stone and who continue to stand aloof.

There is nothing reprehensible in the way mathematicians have worked. Any creative artist knows that criticism before a work is fairly complete is ruinous. Only after the work is far enough along to be offered to the public is criticism relevant—when it cannot cause the artist to spoil his conception. As one of my artistic friends says, it takes two artists to paint a picture: the first to do the brushwork, and the second to shoot him when the painting is finished.

The critics of mathematics have been mathematicians almost without exception. The one reputable exception is G. Berkeley (1685–1753), Bishop of Cloyne, who showed that he knew what he was talking about when he acutely criticized the more bigoted Newtonians. In general the matters in dispute lie far below the surface and are not likely to be observed by any but mathematicians as they go about their business.

In passing, let us remember that Berkeley's specific criticisms were not met until the second half of the nineteenth century, when Weierstrass drove out of analysis the 'infinitesimals,' or 'infinitely small quantities,' of the Newtonians, to which Berkeley had so vigorously objected.

An anecdote concerning the arithmetically minded Kronecker foreshadows one phase of some modern objections to certain kinds of mathematical reasoning. When everyone was congratulating Lindemann in 1882 over his proof that π is transcendental [11.6], Kronecker said, "Of what value is your beautiful proof, since irrational numbers *do not exist?*" Here Kronecker incidentally denied the 'existence' of π, and he was less of a radical at that than some of his successors.

What are the points here? There are several. One which disturbs mathematicians is this very question of what is meant by mathematical existence. We know—or used to think we knew—that with sufficient diligence (and stupidity) $\pi = 3.1415926 \ldots$ could be calculated to an *indefinitely great number of decimals.* Indefinitely great? Not exactly; for who could ever do it? In what sense then, if any, does π 'exist' as an infinite, non-repeating decimal?

I trust that I have not made this sound like a foolish quibble, for it is anything but that.

Kronecker insisted that *unless we can give a definite means of constructing the mathematical things about which we talk and think we are reasoning, we are talking nonsense and not reasoning at all.* At one stroke he denied the validity of all the great work of the mathematical analysts on the infinite. To him it was worse than meaningless; it was useless.

There are those, including many physicists, who say Kronecker was right, and they cannot be silenced by an affectation of superiority on the part of those who believe otherwise. There are equally strong men on both sides of the entire controversy. Arrogance, intellectual or other, has no place in mathematics.

Progress in this direction has been made by meeting Kronecker's objection step by step where it is important to do so, and actually exhibiting finite constructions for things that are used in mathematical arguments. As an example I may mention the constructive proof, after Brouwer, of the fundamental theorem of algebra [5.7]. Even Gauss in one of his proofs overlooked the critical step of proving that a polynomial is a continuous function, yet this attempted proof is still frequently cited by historians of mathematics as the first rigorous proof devised. It would not be accepted today. It is impossible, of course, to meet fully any demand for a construction of an actual infinite; here we have to be content with exhibiting a process which, if carried out, would produce the required thing to any prescribed degree of accuracy.

Having mentioned π ($= 3.1415926 \ldots$) and Brouwer, I may as well give one of his 'counter-existence' examples in support of Kronecker. Suppose someone asserts that somewhere in the decimal development of π the sequence of digits 123456789 occurs. Is this assertion true? Is it false?

Brouwer admits neither possibility because there is no known method of deciding. This kind of skepticism exasperates one school of mathematicians, mostly respectable classicists. Like some other conservatives they seem to believe in Platonic Ideas, although they would indignantly deny the imputation of mysticism. But to continue the example, suppose that some indefatigable calculator does turn up 123456789. Then it may be asked, will this sequence occur again, and if so, where and how often? Suppose n_1, n_2, . . . mark the successive digital places where this sequence appears. Is the series $1/n_1 + 1/n_2 + \cdots$ convergent, or is it divergent? Until someone actually produces n_1, n_2, . . . , the question about the series is unanswerable. The assertion that the series is convergent is neither true nor false. Brouwer (1907, 1912), following Kronecker, demanded constructions for mathematical 'entities' whose 'existence' is purportedly proved without giving any method for exhibiting the 'entities' in a *finite* number of humanly performable operations. As a mere matter of instinct it seems reasonable to suppose that nonconstructive and constructive existence proofs should have different weights in mathematical arguments.

To isolate the essential thing in Brouwer's objections to some traditional modes of mathematical reasoning, I recall that all such reasoning well into the twentieth century was based in part on the classical logic of Aristotle, in particular on his law of the excluded middle [2.2]—some would say the excluded muddle. That law, like some of Euclid's postulates for elementary geometry, was probably an abstraction from sensory experience of one kind or another, specifically in Aristotle's case with reference to *finite* collections of things like heaps of pebbles. What grounds would there be for inferring that experience at this finite level could be consistently extrapolated to the infinite, for which it was not designed, without the possibility of engendering contradictions? Yet until Brouwer suggested that some of the paradoxes of mathematical reasoning might be deeply rooted in an uncritical extension to the infinite of a logic [2.2] devised for the finite, nobody had questioned the universal applicability of Aristotelian logic. Like so many innovators, Brouwer got a warmer welcome from his contemporaries than he had anticipated. But he took off the gloves, as it were, and battered his opponents and detractors, including the titleholder Hilbert, with his bare fists (and caustic tongue) until they either admitted they were beaten or retired from the ring to sulk.

I must warn the reader that the foregoing is an exceedingly crude description of extremely subtle difficulties, and that part, if not all, of it would be regarded as sheer nonsense by one school of mathematical thought. I can only suggest these profound problems and pass on to others, treating them equally sketchily.

The use of words alone in all these discussions is a treacherous proceeding. This also affects much of the technical literature on these disputes. Some mathematicians feel that if the ideas considered can not be adequately expressed in some appropriate symbolism, they are too dangerous to be handled. The history of philosophy is a sufficient warning.

❖ 20.2 ❖ *A Great Illusion*

We have seen [7.2] how Descartes and his successors reduced geometry to a matter of relations between numbers. We might also have seen how Weierstrass imagined the continuum of real numbers [4.2]—corresponding to all the points on a line segment—as a swarm of sequences of real numbers. For example, $\sqrt{2}$ is the limit of a sequence whose initial terms in decimal fractions are 1, 1.4, 1.41, 1.412,

. . . , and there are definite algorithms (such as continued fractions) for exhibiting $\sqrt{2}$ *explicitly* to any prescribed degree of approximation, as the decimal representation does not. But all this detail is inconsequential here. The only thing of any deep significance is that pointed out in 1898 by Hilbert. At that time he had already done some of his greatest work. Fresh from all the triumphs of his early maturity, what did he consider one of the most important, if not *the* most important, problem of 1898 for the mathematicians of the oncoming generations? Simply to prove the consistency of common arithmetic as that elementary domain of mathematics had been understood ever since the time of Euclid. Were that done, the rest would follow— presumably. For analytic geometry is reduced to numbers; the continuum likewise; hence also mathematical analysis. Possible and probable struggles with infinities of various orders as arithmetic was transcended were temporarily ignored. For it seemed evident that unless common arithmetic was first clarified and firmly founded, further progress was likely to be difficult if not impossible.

Hilbert's project of proving the consistency of arithmetic turned out to be much harder than he had anticipated. For one thing, the mere definition of a number as a class of classes similar to a given class, as with Frege and Russell [19.4], revealed deep-lying paradoxes. These diverted Hilbert's original project to another, broader and more fundamental, of reconstructing mathematical reasoning so as to get rid of paradox or at least to obviate it. Thus originated the second and more ambitious part of his program: to prove the consistency of mathematical analysis. But to do this it was necessary to go down to mathematical and logical bedrock and critically revalue the implicitly and explicitly accepted logical reasoning [2.2] of two thousand years or more. Further, classical logic had long been inadequate for mathematics. Hilbert attempted a consistency proof of analysis as the supreme effort of his great career.

Like all good mathematicians, Hilbert distrusted purely verbal arguments. Accordingly, beginning about 1925, he proposed that mathematicians temporarily forget the 'meanings' of their elaborate game with symbols and concentrate their attention *on the game itself and its permissible moves*. What were mathematicians *actually doing?* They were "making meaningless marks on paper" [3.3] and shifting these marks about in accordance with certain tacitly assumed or explicitly formulated rules of play. Hilbert and his pupils attempted to make these rules fully explicit and precise. It was somewhat as if people had been playing chess blindfold for two thousand years, moving the pieces by instinct without grasping the few simple rules of the game, such as that the queen—like the Queen of the Sciences—can move with a certain degree of freedom. The permissible moves of the mathematical game when isolated turned out to be quite few and extremely, even deceptively, simple. For example, for propositions [5.2] p,q, the scheme or move

$$p$$
$$p \Rightarrow q$$
$$q$$

symbolizes that if p is asserted and if p implies q, then q can be asserted—in the relevant context. So wherever p and $p \Rightarrow q$ occur successively we can put q. This may seem trivial until we apply it to a situation in which p,q are complicated compound propositions mixed up in a mathematical argument with several other of the equally simple permissible moves of the too easy game. By refining and elaborating this *pure formalism* of mathematical proof Hilbert hoped to prove that the traditional procedures of classical mathematics, such as those of arithmetic and analysis, *can never* eventuate in a contradiction such as '*A* is equal to *B*,' and '*A* is not equal to *B*.'

Hilbert's theory of proof was to have been the climax of the postulational method in mathematics to which he had contributed at least one masterpiece (on the foundations of geometry) and in which he had been one of the boldest pioneers. But the day of final solutions of the universe and all-inclusive theories of this, that, or the other, whether in philosophy or history or mathematics or logic or science, seems to have definitely ended with the eighteenth century. The toilers of the twentieth century are content to chip off a fragment at a time, leaving to their sanguine successors the illusory achievement of a possibly unrealizable perfection. In this respect Hilbert had missed the spirit of his piecemeal age, although his intended masterpiece might have lasted and been admired for a century had not a keener logician than he demolished it. In the preface to the second volume of his intended masterpiece Hilbert says that his 'theory of proof' had ended in a fiasco—his own word. We shall now see very briefly why he said so.

❖ 20.3 ❖ *From Hilbert to Gödel*

I shall note only the capital result that changed our whole conception of mathematical reasoning. It has been called the most significant advance in logic since Aristotle [2.2]. I shall quote Weyl. With Brouwer, Weyl was one of the pioneers in the modern philosophy of mathematics. His appraisal (1946) of the situation is therefore of special interest.[1]

[1] From *Mathematics and Logic*, the American Mathematical Monthly, vol. 53, 1946, pp. 1–13. Most pure mathematicians and logicians are so accustomed to thinking of K. Gödel as having been born into their respective tribes that it may surprise some of them to know that for three years he studied physics at the University of Brno. In 1949 he published a paper, *An example of a new type of cos-*

It is likely that all mathematicians ultimately would have accepted Hilbert's approach had he been able to carry it out successfully. The first steps were inspiring and promising. But then Gödel dealt it a terrific blow (1931), from which it has not yet recovered. Gödel enumerated the symbols, formulas, and sequences of formulas in Hilbert's formalism in a certain way, and thus transformed the assertion of consistency into an arithmetic proposition. He could show that this proposition can neither be proved nor disproved within the formalism. This can mean only two things: either the reasoning by which a proof of consistency is given must contain some argument that has no formal counterpart within the system, *i.e.*, we have not succeeded in completely formalizing the procedure of mathematical induction; or hope for a strictly 'finitistic' proof of consistency must be given up altogether. When G. Gentzen [1936] finally succeeded in proving the consistency of arithmetic he trespassed those limits indeed by claiming as evident a type of reasoning that penetrates into Cantor's 'second class of ordinal numbers.' [So this purported proof misses Kronecker's demand by an infinity of universes.]

From this history one thing should be clear: we are less certain than ever about the ultimate foundations of (logic and) mathematics. Like everybody and everything in the world today, we have our 'crisis.' We have had it for nearly fifty years. Outwardly it does not seem to hamper our daily work, and yet I for one confess that it has had a considerable practical influence on my mathematical life: it directed my interests to fields I considered relatively 'safe,' and has been a constant drain on the enthusiasm and determination with which I pursued my research work. This experience is probably shared by other mathematicians who are not indifferent to what their scientific endeavors mean in the context of man's whole caring and knowing, suffering and creative existence in the world.

❖ 20.4 ❖ *To Our Successors*

After the splendid achievements of both pure and applied mathematics in the nineteenth and twentieth centuries, it

---

mological solution of Einstein's field equations—quite a long way from pure mathematical logic.

seems ungracious to close on a note of doubt. The senti-
ments of creative mathematicians cannot be disregarded.
Surely their feeling for what is valid in mathematics should
not be ignored. Almost without exception these men feel
this about the past and probable future of their beloved
mathematics: not *all* those giants of the past can have been
fools *all* the time, and we may rest assured that wiser shall
come after them.

Wisdom was not born with us, nor will it perish when we
descend into the shadows with a regretful backward glance
that other eyes than ours are already lit by the dawn of a
new and sounder mathematics, and one that is closer than
the old to human capacity and human needs.

Index

A

Abel, N. H., xiii, 12, 184, 355, 397
Absolute value, 158–161
Absorption, law of, 60, 99
Abstraction, 56, 158–163, 263, 265–266, 269, 271, 287
  and prediction, 252–271
  successive, 271
Acceleration, 210, 213, 325–327
Action, 347, 350
  at a distance, 371
  least, 346, 348
Adams, J. C., xiii, 287, 293, 295–296
Addition, 67–68, 105
  algebraic, 31
  (*See also* Sums)
Adjunction, 86
  algebraic, 87
  simple, 86
  transcendental, 86
Aerodynamics, 351
Aesop, xiii, 158
Airy, G. B., xiii, 293
Alexander the Great, xiii, 272
Alexander, J. W., xiii, 155
Algebra, 7, 9, 22, 56, 95, 101, 131, 140, 175, 193, 221
  abstract, 15, 75, 158, 185, 222, 253
    modern, 84, 222
  associative, 93
    linear, 54, 89, 93, 183
  boolean, 59–60, 76
  of Cayley, 159
  of classes, 99
  classification of, 91
  common, 29–34, 45
    changing rules of, 34–37

Algebra, common, of complex numbers, 72
  postulates for, 30, 351
  realizations of, 45–46
  division, 88
  fundamental theorem, 87, 240, 409
  of invariants, 183
  lattice, 99
  Lie, 183
  linear, 89, 91, 93
  of logic, 55–66, 76
  of mechanics, 92
  modern, 91, 119, 232
  non-associative, 93
  non-commutative, 159
  of rotations, 92
  semi-simple, 91
  simple, 91
  of structures, 55
  tensor, 113
  vector, 37
Algorithm(s), 412
  Euclidean, 243–244
Alice in Wonderland, by Carroll, 361
Almanac, nautical, 293
America, 293, 294
American Mathematical Monthly, 239, 288
American Mathematical Society, 228
  Bulletin, 230
Analysers, harmonic, 369
Analysis, 7, 101, 221, 234–235, 252, 253, 287, 301–302, 305, 351, 379, 402
  abstract, 188
  combinatorial, 374
  diophantine, 236–239
  Fourier, 2

417

Correction to lines 8-10 of page 232.

A line was dropped by the compositor in the original text. We are indebted to Professor Underwood Dudley for the following restoration of the full text, which is taken from Fermat's original:

"If an arbitrarily chosen prime of the form 4n + 1 is not a sum of two squares [I prove that] there will be another of the same nature, less than the one chosen, and [therefore] next a third still less, and so on. Making an infinite descent in this way we finally arrive at the number 5, the least of all of the numbers of this kind."